INTEGRATED HUMAN-MACHINE INTELLIGENCE

INTEGRATED HUMAN-MACHINE INTELLIGENCE

BEYOND ARTIFICIAL INTELLIGENCE

WEI LIU

Professor of AI College, Beijing University of Posts and Telecommunications, Beijing, China

ELSEVIER

清華大學出版社
TSINGHUA UNIVERSITY PRESS

Elsevier
Radarweg 29, PO Box 211, 1000 AE Amsterdam, Netherlands
The Boulevard, Langford Lane, Kidlington, Oxford OX5 1GB, United Kingdom
50 Hampshire Street, 5th Floor, Cambridge, MA 02139, United States

ISBN: 978-0-323-99562-7

For information on all Elsevier publications visit our website at https://www.elsevier.com/books-and-journals

Publisher: Glyn Jones
Editorial Project Manager: Naomi Robertson
Production Project Manager: Punithavathy Govindaradjane
Cover Designer: Mark Rogers

Typeset by TNQ Technologies

Dedication

I would like to dedicate this book to my dear mother, Minhua Zhu and to my family.

Contents

About the author

Wei Liu graduated from Beijing University of Aeronautics and Astronautics and received a PhD Degree in Man-Machine-Environment System Engineering. His main research fields include human-computer integration intelligence, cognitive engineering, human-computer environment system engineering, military intelligent application, future situation awareness mode and behavior analysis/prediction technology, etc. He is author of the book *Integrated Human-Machine Intelligence* (2023, Elsevier). He has also has published dozens of articles in various academic journals. He has participated in multilateral security forums such as China-USA AI Dialogue, Beijing Xiangshan Forum, The ninth World Peace Forum, etc.

Preface

Why would I write the book *Integrated Human-Machine Intelligence: Beyond Artificial Intelligence*? Simply put, it starts with skepticism...

Since I returned home from a visit to Cambridge University in October 2013, I have become more and more uneasy about the new generation of AI wave represented by machine learning, not to completely deny the role of machine learning (its significant role is evident to everyone so I won't list it here), but to worry about its hindrance and stumbling on the future of intelligence. Just as Michael I Jordan, one of the leaders of machine learning, said, "I hate to call machine learning AI, for the former is only a part of AI, but IA (Intelligent Augment) may be more suitable for the development of today's intelligent technology... People cannot expect too much from the so-called AI, which is only to build a 'smart' automation system." So how big is the gap between the current actual ability of artificial intelligence and the intelligence that people really expect? In what way will future wisdom be presented in people's lives? Is the origin of artificial intelligence really the same as what many experts and scholars talk about and write in their speeches and academic papers? ... With this curiosity and doubts, I wrote down some things I saw, heard, read, learned, studied, thought, understood, and found both at home and abroad. During this period, I received much help and inspiration from many teachers and friends, which helped to initially form this book and lay the foundation for the next book. The following is the beginning of my doubts. I will be grateful for any of your generous advice and your insights.

Many things and objects, which are of facts, valuation, and sometimes responsibility, often appear simultaneously in the following four domains: physics, information, cognition, and sociology. In contrast, the data input, reasoning, and decision-making of facts in the human–machine hybrid may change little, while the process of information with value and knowledge cognition is more flexible. How to figure out the combination mechanism of facts and value will become the critical problem of human–machine hybrid intelligence, which is the reason why artificial intelligence is "unintelligent," for it has (visible) facts but no (useable) value.

Existing artificial intelligence is basically formalized factual calculation and lacks intentional value calculation, so it only solves a tiny part of the intelligence problem—part of the automation, which is far away from the expected intelligence... This is because artificial intelligence or automation only solves external problems of input, processing, and output, but ignores more important internal input, processing, and output problems.

The forms of input can be divided into internal and external forms, among which the eyes, ears, nose, tongue, and body are forms of external input, while thinking, reflection, understanding, and meaning are forms of internal input. Language is a form that combines internal and external input.

Processing forms are also divided into two groups: internal and external processing. Bottom-up (rationality) calculation is an external processing, while top-down (non-rationality) suanji (a Chinese term meaning internal processing) and jisuanji (a Chinese term meaning that calculation is combined with suanji) is a processing form integrating both internal and external processing.

Similarly, output is divided into internal and external output, among which logical fact-based decision-making is an external form of output, while intuitive (illogical) value decision-making is an internal output form, and fact-value mixed decision-making is a both internal and external compatible output form.

External input, processing, and output can be collectively described as factual formalization, and internal input, processing, and output can be collectively called valuation intentionalization. Thus, the essence of human—machine hybrid intelligence is essentially a community of tangible facts and values.

Formalized facts calculate approximate linear relationships of various parameter variables, while the essence of intentional the value of suanji is not to deal with the relationship between each parameter variable element, but to deal with the interaction of various relationships. Intentionalized value suanji is equivalent to the renewed arbitrary segmentation, reorganization, free arrangement, and a combination of heterogeneous formalized factual calculation.

Any intelligent system needs error correction. Jisuanji, namely machine's calculation + human's calculation, can be designed as a system of error correction mechanism to realize Wittgenstein's nonfamily similarity accommodate compensation verification. True intelligence or artificial intelligence is not what abstract mathematical systems can achieve. Mathematics, which can only realize functions but not abilities, is just a tool. Only humans can produce actual abilities. Therefore, artificial intelligence is the product of the interaction of human beings, machines, environments, and systems. In the future, the intelligent system should also be the insight results of suanji made by human beings combined with machine calculation and environmental changes. Automation is deterministic input, programmable processing, and deterministic output; artificial intelligence is partially deterministic input, programmable processing, and partially deterministic output; while intelligence is the input of uncertainty, unprogrammable processing, and the output of uncertainty. The difference between artificial intelligence (including automation) and intelligence is that the former is a function, and the other is a kind of ability. Many people expect ability, not function, to achieve the purpose of intelligent ability through the means of human intelligence function. This is the contradiction between the ideal and reality, and it is also the disappointment of people to regard function as ability. Function refers to the beneficial role of things or methods, the nonactive efficiency of nonliving bodies, such as hammers, cars, robots, etc. Ability is the comprehensive quality of the main initiative embodied in the completion of a goal or a task. It is also a measure of the level of exploration, cognition, and transformation of living objects about their nature, such as human ability, animal and plant reproductive ability, etc.

In fact, the intelligent mechanism in artificial intelligence is communication and interaction. A good fighter pilot and a good manager are often people who break the rules of communication and interaction. They are not people without rules, but they have their own rules which make them wander in the interactive interface. Communication is a shallow touch and mixing, and interaction is deep penetration and integration. Good interconnection and interaction requires emotional injection, such as sympathy, empathy, courage, love, curiosity, responsibility, etc. Therefore, many interfaces only communicate without interaction, and many humans and machines do not interact. People often use another Chinese character together with the character "jiao" (a character meaning "communication") to refer to friendship, interconnection, negotiation, exchange, entangle, trade, and so on, among which the words after "jiao" have the meaning of "interaction." Only from the system's perspective can we gain insight into the essence and authenticity of complexity, intelligence, and cyberspace. Only from the system's perspective can we understand the algorithm model and its internal and external interconnection and understand the integration intelligence of people, machines, and the environment.

No matter how powerful AlphaGo is, it is always a series of automatic computing and responses to meet its established goals (we occupy more space than our opponents) with the set rules. AlphaGo cannot give answers to those questions which do not have established rules, lack setting goals, are not suitable for mathematical calculation, or those with the results simply based on quantitative advantages (these questions should be treated differently from the "things," for the former includes confused and uncertain emotions, philosophy, artistic conception, yearning, etc.). AI learning can support the writing of poems, because those poems are formed by words selected from many existing poems according to the designer's understanding and then combined according to specific rules, which are always "beautiful". These words are not proof that machines can express their feelings in words. In fact, human emotions are difficult to express by logical rules, such as showing weakness to gain the initiative or apologizing when they are aware of offending others. AlphaGo can never get those feelings.

In this book, we will discuss and study the main issues involved in human—-machine hybrid intelligence, which is an essential direction in artificial intelligence and intelligence, such as the ideological differences between the East and the West, dynamic representation, autonomy, formality and intention, nonexistence, facts and values, calculation and suanji, linearity and nonlinearity, perceptuality and rationality, the relationship between state, trend, sense, percept, and information/automation/intelligence, etc.

We will discuss some questions from artificial intelligence to human—machine hybrid, from rationality to sensibility, from the West to the East, from the past to the future. If the last book, "*From Cambridge to Beijing,*" is a visible journey, then this book is an invisible way of mind. Maybe life is composed of a lot of these "visible" and "invisible" things. Once upon a time, I stood at the ruins of the camp castle built by the Romans and watched the Cam River slowly flowing through the famous bridge, feeling about the origin of Cambridge. Until now, looking back on the past, I can feel my dreams still haunt the mountains, waters, and bridges, but through the shadows of the past swords and bridges, and I can still faintly see a round of red sun rising in the east. The beautiful glow puts a layer of magic and exclamation on the statue of Confucius in the back garden of Claire College and the statue of Taylors next to the ancient market in the bustling city center of the bells.

CHAPTER 1

The essence of intelligence

Contents

2020 is destined to be an unforgettable year for human beings. In addition to the COVID-19 epidemic sweeping the world, there is also a strange phenomenon that the COVID-19 infection and the death toll of the United States rank first in the world though its intelligent conditions such as human parts, software, hardware, and other intelligent conditions are the first in the world. It is not difficult to notice that the United States situation not only falls in the first place but also appears as a large-scale imbalance in the system of man + machine + environment. Therefore, the wrestling between China and the United States not only focuses on the leading advantage of each or all individual artificial intelligence, but also the integration of intelligence in the integration of human, machine, and environmental systems. The following analyzes and discusses these intelligence problems.

1. The source of intelligence in the east and the west

Why do so many people question future artificial intelligence? One reason is that current artificial intelligence is far from meeting people's expectations, and the AI we see now is automation or advanced automation in a sense. So what is the difference between intelligence and automation? Automation refers to a fixed input and expected output. For example, many production lines are automatic production lines. By the contrary, intelligence is not the same as automation, for the input can be either fixed

Integrated Human-Machine Intelligence
ISBN 978-0-323-99562-7
https://doi.org/10.1016/B978-0-323-99562-7.00001-2

or not fixed, while the output is bound to be unpredictable even some unexpected things. And this is intelligence. What is intelligence? There are two theories. First, Mencius wrote that "The sense of right and wrong is wisdom" You can be conscious, but not necessarily wise. Consciousness has nothing to do with the sense of right and wrong, but wisdom is to know what is right or wrong and understand what is ethics. According to our research, ethics and human intelligence should be very close. What is ethics? From the perspective of ancient Greece, the principle of classification is ethics. We should notice that the essence of intelligence is also to classify right and wrong. This topic is very interesting. You can check the relevant literature, especially the ancient Greek literature, and you may find that they regard ethics as the principle of classification and distinguishing whether people have morality and ethics. Following the right path to get what you want is called morality. If you do not follow the right path, there will be no morality. What is benevolence? Benevolence means people (the Chinese character of "benevolence" is the homophone of the Chinese character of "people"). And what is righteousness? Righteousness is what "should" be done. Mencius specially emphasizes righteousness and justice. This "should" may be the common intersection of Eastern and Western intelligence. The difference of "intelligence" between the East and the West relies on that one emphasizes analogies and metaphors while the other one emphasizes induction and deductions.

Fig. 1.1 lists some celebrities related to intelligence, Easterners are below and Westerners are above, among which the first person is Thales, the

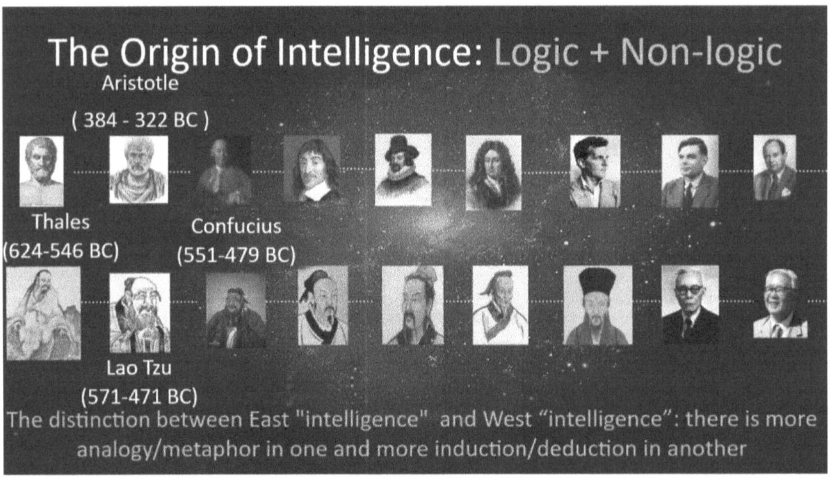

Figure 1.1 Eastern and western celebrities related to intelligence.

ancestor of Western science and philosophy, who was the first to see the world and the total solar eclipse from a rational perspective. The third is Hume, who many AI people do not know about, who was the root of the origin of intelligent philosophy. Hume's famous question is whether value can be derived from facts. This is Hume's important point of view, and it may be the breakthrough point of strong artificial intelligence in the future, and it is also the key point of man-machine integration intelligence. The sixth is Leibniz, the source of intelligent science, who first put forward "universal writing" and "rational calculus" On the basis of above two words, Frege put forward the term "meaning" in analytical philosophy. Later, Boolean appeared, and Boolean algebra was evolved from Leibniz's thought too, following by that Turing and Von Neumann all extended from this. The real origin of technology was Leibniz, and then everyone knows Turing if Leibniz is not so well-known in the field. In fact, Turing's teacher and friend was Wittgenstein (the seventh in the picture, also the one rarely mentioned by AI researchers), who was a very powerful man. There are several special things about him, among which the first one is that his family member was the iron and steel king of Europe, but he loved analytical philosophy. The second special point is that his student and friend was Turing, whose discussion and argument with him gave Turing a lot of good intelligence and philosophy, but many people do not know him, which is a great pity, and it is also the regret of the artificial intelligence community. It is impossible not to mention him. There were two books in his life. The first book was the *Theory of Logical Philosophy*, which was translated into Chinese with no more than 100 pages. In this book, human language is the starting point to pry open the relationship between people and people, and between people and the world. Standardized social language is very important, that is, speaking one sentence after another according to regular grammar, so one can understand the world. When he was in his 40s, he returned to Cambridge University and wrote some manuscripts, which were sorted out into another book called *Philosophical Studies* later by one of his female students, Anskom. This book was not thick either and translated into about 200 pages of Chinese. He negatively inherited the idea of the first book, believing that the most important starting point for a real understanding of human intelligence is naturalized and daily language. For example, there is no grammatical dialogue between people in the bazaar. These two books are the philosophical basis of weak artificial intelligence and strong artificial intelligence. The first book is related to logical philosophy and the second book deals with philosophical research without logic. The real strong artificial intelligence is certainly not all logic. With only logic, it is automation, which is regularization (or add some seasoning, statistical probability). What holds the hearts of the people and human wisdom are the things that are not rational. Hume, Leibniz, and Wittgenstein were the real sources of intelligence.

The Eastern thought starts from "*The Book of Changes*" which is quite outstanding. China has only Eastern thought, but the West does not recognize that there is philosophy in the East. They hold the opinion that there is little logical but only fruitful consultant content in Eastern thought. The first writer of the book was Fuxi, who began to write "*The Book of Changes*" when he saw the changes of the four seasons. The second author of the book was Emperor Wen of Zhou Dynasty, who wrote "*The Book of Changes*" when he was detained. Emperor Wen of Zhou put social management as well as nature into the book. The third author of "*The Book of Changes*" was Confucius, who put the ethics among people into the book, which is the comprehensive integration of the three writer's thoughts. There are three words in "*The Book of Changes*" that are the core of human wisdom. The first word is "Zhi Ji" which is seeing the signs of the development of things. The second word is "Qu Shi" which means seizing the opportunity in time. The third word is "Bian Tong" which means it changes according to the circumstances and changes from time to time. Are there any of these things in the current smart products? No. Those things, so-called merging signs, so-called seizing the opportunity, and so-called accommodation to circumstances, are mostly automation. The second book is *Tao Te Ching* written by Lao Tzu. The first sentence in this book is probably the most important sentence in intelligence, "The truth can be known, but it may be not the well-known truth. Things may be named, but names are not the things." Lao Tzu's "name" involves the first stage of intelligence: the input representation stage. In the West, it is called representation, which means expression, representation, and characterization. There are thousands of representations of a thing. A world in a wildflower, and a bodhi in a leaf. One thing can be represented in many aspects by people, while it is very awfully represented in the knowledge graph. The objects, attributes, and relations of the knowledge graph are all fixed. It is a tagged world. When we could see an unfixed knowledge graph can not be expected, because the existing representation methods can not solve the representation problem. The "Tao" in *Tao Te Ching* contains algorithms, which are not simply mathematical calculation methods but include people's nonlogical calculation methods. Tao is a unique intuitive thing of human beings. There is calculation as well as suanji in suanji, which can pass through nonfamily similarity. In contrast, there are rules in calculation, and it is difficult to integrate unstructured data, nonlinear algorithms, non-faceted reasoning, nonsexual judgment, and nonsystematic decision-making.

Here we briefly introduce the last two gentlemen from the East. Mr. Jin Yuelin, who studied Western philosophy and mathematical logic very well, was one of the pioneers who first introduced Western logic to our country. The second one is Mr. Hua Luogeng. Without Hua Luogeng, there would be no Chinese calculation and Chinese artificial intelligence. He first proposed to track the world's new technology to do computers. Hua Luogeng went to the United States to study advanced technology, and he urged China to develop computers while finding the computing power of computers when he saw Von Neumann make computers. After returning to China, Hua Luogeng assembled some people at Tsinghua University and set up the Institute of Computing Technology of the Chinese Academy of Sciences. The earliest calculation in China is still in Qufu Normal University, and the old machine still exists there. You can stop by and have a look at it when you visit Confucius House and Confucius Temple. If you want to chase the roots of Chinese computers and artificial intelligence, it can be traced back to Hua Luogeng. Later, some old scholars of the Chinese Academy of Social Sciences were tracking the ideas of Wiener cybernetics in the Soviet Union and Europe and the United States, which ignited the spark of artificial intelligence in China. Later, some calculation and automation teachers gradually held various meetings and set up groups in the field of artificial intelligence. In doing knowledge, they must dig their roots and find the real root before they can grow up. Otherwise, what is the scientific origin of intelligence? What is the philosophical origin of intelligence? What is the origin of technology? It was not 1956. The thing that happened in 1956 was an appearance. According to the textual research of a friend, the concept of AI was not put forward by a few people at Dartmouth College, but by a British mathematician who wrote to those people in Dartmouth College. Chinese scholars are now a bit confused about it, and they have not yet learned how to trace its source. Only by tracing to the most fundamental place can the real intelligence of China be produced. Especially for the very good oriental ideas and a lot of intelligent content in ancient China, it is far from enough to dig. We can only see some people make goods and systems, that is not making the intelligence but making money. Only when it is really not for the simple purpose of making money can China's artificial intelligence thrive. Otherwise, it will be very difficult to do anything.

Artificial intelligence originates from the formal logic framework, while human intelligence is rooted in the dialectical thinking system, and man-machine fusion intelligence basically lies in the combination of logic and nonlogic. No matter if it is right or wrong, only the positive and negative matters, not only superposition, but also entanglement.

2. The generation of intelligence

The mechanism of intelligence generation has always been the focus of attention in many fields, involving a wide range and depth of fields. Preliminary carding may be related to the most basic questions: the mechanism of cognitive generation, the mechanism of knowledge generation, the mechanism of meaning generation, the mechanism of emotion generation, the mechanism of situation generation, and even the basic question of philosophy. For example, is the origin of the world material or conscious? Who am I? Where did I come from? Where am I going? What is the means of knowing the world? Is language the key to deciphering human intelligence? What is the relationship between mind and phenomenon?

History has proved that these problems are far from being solved by a few seminars held by mathematicians, philosophers, physicists, calculation experts, automation experts, sociologists, psychologists, and linguists. The intellectual ideas of pioneer masters such as Leibniz, Wittgenstein, Einstein, Schrodinger, Turing, Wiener, Shannon, Bertalanffy, Von Neumann, Simon, Minsky, and Hinton did not change a lot as expected in the academy. This is a bit like the mechanism of love. There are thousands of explanations and understandings of the stories of Romeo and Juliet, 1000 pairs of Xuxian and the Snake White. For human beings, this is an eternal topic, and it is the dream lover and ideal home pursued by lovers for generations. In any case, there is still practical value in the truth that "if there is no human, there will be no intelligence, and there will be no artificial intelligence."

For many reasons, people often associate intelligence with science and technology, which is referred to as intelligent technology for short. It is wrong. Intelligence predates technology, and it appeared when people used stones, sticks, and fire to survive, and there was no scientific technology at that time.

There is no doubt that after the creation of science and technology, intelligence has played a very important role in the development and evolution of intelligence itself and has especially greatly changed people's food, clothing, housing, transportation, and spiritual world. Scientific research adopts observable, measurable, and provable methods. This means that human beings can observe and measure a phenomenon or a problem, and then use mathematical tools to formally describe it as strict and accurate knowledge, and then find regular explanations or conclusions on specific natural and social phenomena or problems and make empiricism or falsification. However, then there appeared the unpredictability of physics, the impossibility of economy, and the incompleteness of mathematics. Slowly, finally, people began to doubt modern science as much as they doubted thousand-year-old theology.

The generating mechanism of intelligence, probably like the three questions of "I" in philosophy (who? Where are you from? Where are you going?), in essence is a cultural problem, and intelligence is also the result of the interaction of many cultures. Among them, Hume's question may be a starting point: Can subjective value be deduced from objective facts? That is, how to derive the "should" question from "being" Almost all intelligent generation will involve subjective purpose and motivation (intentional or unintentional), and it will be related to the changes of objective facts in the situation. The key to answering Hume's question is the cracking of various mechanisms of explicit and implicit analogy (for example, the subconscious is the implicit analogy). In this regard, Hou Shida made a good reflection in his book *Appearance and Essence* on the question, but there are still some problems to be discussed. For example, the appearance and essence of things are often nested with each other, inconsistent, specious, and so on. In fact, the process of human understanding is the process of finding the value "should" in the fact "being" There is a dictionary to interpret it as "to know the meaning of something" The "know" is principal, and the "meaning" is personalized. Therefore, strictly speaking, understanding is self-righteous, while intelligence is seeking truth from facts. Intelligence is not divided into fields, but it can be transferred across domains, so military intelligence should accurately be intelligent military, and intelligent agriculture, intelligent transportation, and intelligent medical treatment are the applications of intelligence in different fields, but they are similar in many basic mechanisms, such as representation at the input, reasoning mechanism in the process of understanding and fusion, decision-making assistance at the output, and so on.

Real intelligence research includes not only game decision-making with incomplete information but also intuitive insight with complete information (such as whether you can assemble Luban lock and Rubik's cube event though all the real materials are given to give). What is the most important representation of intelligence is where the key point of decision-making is, what the focus is, and how to use data, information, and experience properly, rather than the pile of CNN, RNN, ANN, DL, RL, Bayes, and Markov. To achieve this, we need to calm down and ask ourselves the following: What is wrong with these existing conventional methods/parameters? What can be formalized? What cannot be formalized? How do we grasp these "key points" and find the key to solving these key problems?

On the surface, Hume's question is a question of subjective and objective connection, that is, whether the universe keeps rolling on orbit (objective law-theory of relativity) and a superior man makes untiring endeavor for advancement (subjective will-outlook on the world) can be transformed into each other. In fact, there is another key point in Hume's question: that is inference, which involves the incompleteness of induction and deduction. And more importantly, the inference can be related to analogical argumentation, especially from the differences between psychological and physical phenomena.

"People find things like memory, reasoning, feeling pleasure and pain in themselves. People don't think that sticks and stones have this experience while other people surely have the experience". The analogical guess of the intentions of others is obviously different from the analogy of physical facts, for the former requires an assumption different from the physical interpretation. So people resort to subjective and objective cross-border analogies. "Other people's behavior is similar to our own in many ways, so we assume that there must be similar reasons" (*Nature of the Heart*, Rosenthal: Oxford University Press, 1991, p. 89). Others behave in the same way as we do, so when we feel depressed (or happy), others will feel equally depressed (or happy). In other words, the similarity of physical behavior should not only be explained by physical and physiologic causality, but also be able to deduce intellectual, conscious, and emotional similarity. This mechanism of empathy is actually the premise and basis for achieving effective dialogue and coordination between people and machines.

If the universe keeps rolling on orbit, does a man have to make untiring endeavor for advancement? It is hard to get such a conclusion in Hume's point of view, but it is a certain case in "*The Book of Changes*" Change involves not only natural order and human society, but also people themselves. This is also the difference between the state, trend, sense, and percept of the East culture and the situation awareness of the West culture.

A clear understanding of things should not be based on the facts or things, but through comparison with the frame of reference constructed by other things. People's cognition of things is generally triggered by multiple frames of reference, including the organic integration of explicit and hidden coordinate systems (these components are included in games such as the Turing test).

Intelligence has the nature of the timeliness, and the intelligence of each generation is different. From a certain point of view, Newton's intelligence is not as good as that of any college student majoring in physics, at least Newton does not know the existence of the theory of relativity. However, Newton transcended his time and illuminated the ignorance and darkness of his contemporaries and previous generations. From the angle of concepts such as Zhiji, Qushi, and Biantong, intelligence is also a kind of art form, of which the foundation is emotion and the philosophy is aesthetics. Analogy, comparison, and similarity are all important ways to make the art of intelligence. In many situations, the enhancement and attenuation of analogy mechanism often means the strength of intelligence.

Recently, Bengio said that deep learning needs to be corrected. He believes that only by going beyond the scope of pattern recognition and knowing more about causality can it make a real revolution in artificial intelligence. In other words, he says, deep learning requires one to start asking why things happen. Cognitive science experiments also show that understanding causality is the basis of human development and intelligence, although it is not clear how human beings form this kind of knowledge.

These views are both right and wrong. It is right because it is half a step more advanced than machine learning: it no longer relies solely on statistical correlation analysis to move forward mechanically. It is wrong because it has not yet stepped out of the traditional thinking of the Western scientific and technologic workers: causality is still regarded as a panacea for scientific and technologic roots. In fact, analogy, comparison, metaphor, and similarity may be the most important ways to achieve creative intelligence (including fitting to generate a variety of new concepts). If we pay attention, we will find that children usually like to say that something is like something, and they always like to compare known things to unknown things from the angle of external state appearance attributes such as shape, color, and size to the essential relationship of internal trends such as time, space, and change. Maybe this is the secret of human cognition. "Cause and effect" is more likely to be just another name for "effect and cause" Whether the apple falls to the ground or Mercury is biased is one such phenomenon. From this point of view, Newton and Einstein are probably subjective idealists. Coincidentally, someone once said to the point, "In essence, mathematics is the knowledge of concepts." Of course, all concepts are related to subjectivity.

Intelligence, including artificial intelligence, is a complex system, in which many things cannot be explained clearly by logical thinking, and there are a large number of nonlinear and nonlogical components, interpretability, lifelong learning, dynamic representation, strong and weak reasoning that all need analogy, but the mechanism of analogy is far from being solved by science and technology alone, especially when it involves emotion, situation, virtual body, and so on. Trying to solve the main core problems of intelligence or artificial intelligence simply with mathematics, especially with modern incomplete mathematics, is tantamount to seeking fish from wood, drawing cakes to satisfy hunger, and fishing for the moon in the water, just like building airplanes and rockets with wooden sticks and stones. The reason is simple: the qualitative real things are still being explored in the distant future, and the quantitative ones can only be automation.

There is no doubt that artificial intelligence is based on mathematics. At the same time, the application of artificial intelligence, however, is inseparable from the people and environment who design, manufacture, use, manage, and maintain it. This system of human, machine, and environment often transcends the shackles and constraints of mathematics, forming a mixed situation of mathematical and nonmathematical fields. How do we deal with this complex situation effectively in time? It puts forward a very difficult and urgent problem for people. Mathematics is to deal with the problem of fact, and people can deal with the problem of value, especially the use of mathematical methods that can better deal with the mixed problem of fact and value.

What is intelligence then? Humans have known a part of intelligence, not all of it.

Intelligence has always been the object of common concern between Eastern and Western civilizations. Mencius said, "The sense of right and wrong is wisdom" Mises said, "The difference between A and non-A is knowledge" The right and wrong can be replaced by "to be or not to be" in the West, and the activity between the two (what "should" means) is intelligence. Westerners prefer philosophy and science to oppose superstition (there are too many gods). The Chinese prefer historical ethics to be people-oriented (there are many people). In fact, there are all these elements in intelligence. Intelligence contains logic, and there are a large number of nonlogical elements at the same time, such as intuition, non-axiom, vagueness, and so on. In intelligence, there is not only the confrontation of logical/ethical paradoxes but also the compromise of logical/ethical paradoxes, which in essence is the representation of unitary reality without losing its meaning with pluralistic possibilities. In short, it is to express the only reality with thousands of possibilities, to express the complexity and the degree of concentration.

Intelligence is the ability of irrelevance, that is, it is the ability to find, analyze, and solve problems that are ostensibly unrelated (intrinsically inextricably related). The index to evaluate intelligence is the complexity of dealing with the problem. The first relationship is not the relationship between intelligence and things, but intelligence is the relationship among things. Intelligence embraces everything into existence and maintains it in existence. In this regard, intelligence itself is relevance regardless of causal connection. In the narrow sense, intelligence has time and space (such as artificial intelligence), which requires adaptive processing of information when resources are limited; in a broad sense, intelligence has no limit in time and space (such as wisdom). Using the methods with infinite material, we achieve the goal; this cross-border with or without time and space is also one of the reasons why intelligence is difficult to define.

Intelligence is the subjective construction of relationship, and it is also a reflection of objective existence and a product of the integration of subjective and objective. The so-called description of the objective existence is actually to sort out and analyze the interactive relationship between the character (machine) environment system clearly. It is really difficult to take the current mathematics as the model, because there are some incomplete elements in it. In this sense, the expectation of a breakthrough in intelligence is actually a desire for new methods and means of description. Although the dual division method of subjective and objective provides a lot of convenience for human beings to explain the world, it also sets up certain obstacles to transform the world. How to break it (for example, to set up a third-body perspective in addition to the subject and object) may be a breakthrough direction in the future.

The current intelligence itself is not a knowledge that can be solved by science, mathematics, philosophy, or the humanities alone. For example, mathematics can accurately describe physical objects, but it is difficult to describe complex processes, especially psychologic, social, and cognitive processes. Therefore, some people use this sentence to describe the limitations of mathematics: "Mathematics can well solve the accuracy of the problem, but it is not easy to solve the direction of the problem." The intelligence of the future is not a problem that can be solved by a future discipline alone. It is essentially a complexity problem, which requires the continuous cross-integration of many fields. Of course, some progress can be made through the efforts of some disciplines, but these advances may cause some invisible retrogression or hindrance, that is, the retrogression of progress.

3. The past and present state of artificial intelligence

The idea of artificial intelligence originated in the 1940s. In 1948, Turing described thinking machines in his paper "Computing Machines and Intelligence," which was regarded as the embryonic form of artificial intelligence and proposed the "Turing Test" In 1956, Marvin Minsky, John McCarthy, Claude Shannon, and other scholars held a seminar on artificial intelligence at Dartmouth College in the United States, which was called the "Dartmouth Conference" and formally established the concept and development goal of artificial intelligence. Research fields consist of propositional reasoning, knowledge representation, natural language processing, machine percept, and machine learning. The development of artificial intelligence for more than 60 years can be summarized into the following main development stages. The first peak period (1956 to 1974) was after Dartmouth Conference, where researchers made some achievements in the fields of search reasoning, natural language, machine translation, and so on. The first trough period was from 1975 to 1980: with the failure of machine translation projects due to the lack of calculation computing power, high calculating complexity, and difficulties in the realization of common sense and reasoning, artificial intelligence began to be widely questioned and criticized. The second peak period was from 1981 to 1987: expert systems with logic rules deduction and problem-solving in specific fields began to prevail, with Japan's "Fifth Generation Calculation Program" as a typical representative. The second trough period was from 1988 to 1993: abstract reasoning was no longer concerned, and intelligent models based on symbolic processing were opposed. The development period was from 1994 to 2012: the emergence of artificial intelligence systems such as "Deep Blue" made people feel the infinite possibility of artificial intelligence again. The explosive period is from 2013 (until now), which is the new generation of information technology, such as machine learning, mobile network, cloud computing, big data, and so on. It has led to changes in the information environment and data foundation, with further acceleration of computing speed and a substantial reduction in cost, promoting the explosive growth of artificial intelligence to a new generation.

Modern artificial intelligence is a bit like primary school students doing homework. It only focuses what is assigned, lacking autonomous/automatic generation of demand tasks, dynamic task planning, and demand contradiction coordination. It is difficult to deal with the contradiction between fast situation awareness and slow situation awareness, but it is not easy to realize the organic interaction of the whole man-machine environment system and the mixed embedding of facts and value elements.

Tracing back to the origin of modern computer science, we should say that it has a close relationship with logic. As we all know, since Russell and Whitehead co-wrote *The Principles of Mathematics*, the study of mathematical logic has sprung up. People even started to expect to build the whole mathematics and even the science building on the basis of logic. Driven by this logicism, it is inevitable to formalize the concept of "being feasible and calculable" In the exploration of the concept of "being feasible and calculable" Church, Godel, and Turing gave completely different but equivalent definitions almost at the same time. Church invented the Lambda calculus, which is used to depict "being doable and computable" Godel proposed "general recursive function" as the definition of "being feasible and computable" Turing defines the concept of "being feasible and calculable" through the description of a device, which was called "Turing machine" by later generations and is the theoretical model of modern calculation that marks the birth of modern computer science.

In the 17th century, logic changed its researching focus, and Leibniz put forward what logic should do. Leibniz aimed to establish a universal language for science, which is ideal and appropriate for science to reflect the nature of the entity in the form of sentences. Leibniz believed that all scientific ideas can be classified as fewer, simple, indecomposable ideas, using them to define all other ideas, and by decomposing and combining ideas, it is possible to make new discoveries, such as the calculus process in mathematics.

Leibniz first discovered the universal meaning of symbols: human reasoning is always carried out in the form of symbols or words. In fact, it is impossible and unreasonable that things themselves, or their ideas, are always clearly identified by the mind, symbols, so they need to be used for the sake of economy. Because each time a geometrist mentions a conic, he will be forced to recall their definitions and the definitions of the terms that make them up, which is not conducive to new discoveries. If a mathematician constantly needs to think about the values of all the tokens and passwords he has written in the process of calculation, it will be difficult for him to complete large-scale computers. Similarly, when a judge reviews the actions, exceptions, and interests of the law, it will not always be possible to make a complete review of all these things thoroughly, which will be huge and not necessary. Therefore, we give names to geometric shapes and symbols to numbers in arithmetic. In algebra all symbols are found as things, either through experience or reasoning. Finally, it can be fully integrated with the symbols of these things, and the symbols mentioned here include words, letters, chemical symbols, astronomical symbols, Chinese characters, and hieroglyphs. It also includes musical symbols, shorthand, arithmetic and algebraic symbols, and any other symbols that people use in the process of thinking. Here, "text" is written, traceable, or carved text. In addition, the more a symbol is able to express the concept it refers to, the more useful it will be. It can be used not only for representation but also for reasoning.

Based on this, Leibniz noticed that all objects can be assigned their words and numbers, so that a language or text can be constructed, which can serve the art of discovery and judgment, just as arithmetic is to numbers and algebra is to quantities. People will inevitably create a letter of the human mind, and all things can be found and judged through the comparison of the letters in the alphabet and the analysis of words made up of letters,.

In Leibniz's insight, there are two important concepts, namely, "universal writing" and "rational calculus" The so-called "universal characters" are not chemical or astronomical symbols, nor are they Chinese characters or ancient Egyptian hieroglyphs, let alone people's daily language. Although people's daily language can be used for reasoning, it is too ambiguous to be used for calculus; that is to say, everyday language cannot detect errors in reasoning through the formation and construction of words in the language. Comparatively speaking, arithmetic and algebraic symbols are the ones that are most similar to "universal characters" In arithmetic and algebraic symbols, reasoning exists in the application of words, and the fallacy of thought is equal to the error of calculation. Universal writing is a kind of letter of human thought, and everything can be found and judged through the connection and the analysis of words.

Another key concept is "rational calculus" where calculus is different from reasoning. It is a calculation or operation, i.e., the relationship produced by transformation of formula made up of one or more words according to some preset rule. Closely related to the concept of calculus is the art of writing, the art of forming and arranging words, by which they represent ideas. In other words, it makes the relationship between words the same as the relationship between thoughts.

It should be mentioned that Leibniz's idea of universal writing and rational calculus is very "idealized and utopian" Leibniz only put forward such an idea, but it has not been fully realized though he himself made some attempts. Therefore, many logicians and philosophers call Leibniz's idea "Leibniz's dream" After that, the work of Hilbert and Godel showed that there was no such perfect language and calculus. Based on this, many scholars asserted that Leibniz's dream had been broken.

In a speech held by the Mathematical Society of London in 1947, Turing expounded some of his views on symbolic logic and philosophy of mathematics, "I hope that digital computers will eventually arouse our considerable interest in symbolic logic and philosophy of mathematics." The communicative language between humans and these machines, which is known as instruction table language, forms a kind of symbolic logic. Machines explain everything we tell them in a fairly precise way, without reservation and without a sense of humor. Humans must accurately convey their meaning to these machines, or there will be trouble. In fact, humans can communicate with these machines in any precise language. In essence, we can communicate with machines with any symbolic logic, as long as the machine is equipped with an instruction table that can explain this symbolic logic. This means that the logical system has a wider range of application than ever before. As for the philosophy of mathematics, the focus of human interest will continue to shift to philosophical problems as the machine itself will do more and more mathematics all by itself.

The Turing machine has two main components: "automata" and "instruction table language" The instruction table language refers to the language that describes the state transition table, which is the language that describes the state transition, reading, writing, and movement of automata. Turing believed that instruction table language is a communicative language between human beings and machines, forming a kind of symbolic logic. The instruction table language here includes the various types of programming languages that people later developed.

We could say that the "automata" and "instruction table language" of the Turing machine are the interpretation of Leibniz's "universal text" and "rational calculus" Turing more successfully realized the Leibniz dream than Frege, Bull, and Russell did. In Turing's scheme, programming language is an implementation of universal text (programming language - ≈ universal text). Automata is a kind of realization of rational calculus (automata ≈ rational calculus).

Turing effectively combines Leibniz's "universal text" with "rational calculus" which brings a kind of "computing turn" to logic. We can say that besides the logic as algebra and the logic as language, Turing opened up a new path for logic: "calculating logic" This kind of examination of logic is essentially a kind of "change of the main body" The "former logic" takes human beings as the main body, and the object of study is human thinking and all kinds of natural languages that represent human thinking. "Calculating logic" is to regard the calculation as the main body of information processing, and to study the processing mode of the calculation and the interactive relationship between the human and the calculation.

The 20 years after the birth of artificial intelligence is the dominant period of logical reasoning. In 1963, Newell, Simon, and some others compiled the mathematical theorem proving program of "logic theory machine" (LT). On this basis, Newell and Simon developed a general problem-solving program (GPS), which opened up a major field of artificial intelligence: "problem-solving" Classical mathematical logic is only mathematical formal logic, which can only meet part of the needs of artificial intelligence.

After that artificial intelligence developed a numeric method to represent and deal with uncertain information to assign a value to each statement or formula in the system to express the uncertainty or certainty of the statement. Some of the representatives are as follows: the subjective Bayesian model put forward by Duda in 1976, the possibility model put forward by Chad in 1978, the incidence calculation model put forward by Bondi in 1984, and empirical models such as hypothetical reasoning, qualitative reasoning, and evidence space theory empirical models.

Inductive logic is the logic of probabilistic reasoning. In artificial intelligence, induction can be regarded as reasoning from individual to general. With the help of this method of induction and analogy, the calculation can call relevant knowledge from the corresponding knowledge base to deal with new problems through the similarity between new and old problems.

Common sense reasoning is a kind of nonmonotonic logic. In other words, people draw some conclusions based on incomplete information. When people get more complete information, they can change or even withdraw the original conclusions. Nonmonotonic logic can deal with reasoning in the case of insufficient information. In the 1980s, Wright's default logic, McCarthy's finite logic, the nonmonotonic logic reasoning system established by McDermott and Doyle, and Moore's self-cognitive logic are all groundbreaking nonmonotonic logic systems. Common sense reasoning is also a kind of inaccurate reasoning that can make mistakes, which is fault-tolerant reasoning.

In addition, multivalued logic and fuzzy logic have also been introduced into artificial intelligence to deal with the reasoning of fuzzy and incomplete information. The three typical systems of multivalued logic are the ternary logic systems of Klein, Lucasivez, and Bockwan. The study of fuzzy logic began with the study of Lukasewicz in the 1920s. In 1972, Zade put forward the principle of relation composition of fuzzy reasoning, and most of the existing fuzzy reasoning methods are the deformation or extension of relation synthesis rules.

From Leibniz's "universal text" and "rational calculus" to Turing's "instructive language" and "automata" from inductive logic to common sense reasoning, multivalued logic, fuzzy logic, and so on, modern artificial intelligence can be regarded as the continuation and development of "universal text" and "rational calculus."

4. Three stages of the development of artificial intelligence

In 1956, dozens of experts spent 2 months having a thorough discussion on artificial intelligence at the meeting held in Dartmouth, in which the concept of artificial intelligence originated. In the following 60 years, the development of artificial intelligence has been progressing continuously, witnessing three obvious stages of development.

4.1 The first stage featuring the inference system

The first stage lasts from 1956 to 1976. In this stage, people think that if the machine can be as reasoning as humans, it is intelligent. The landmark event in this stage is the appearance of an inference system that can prove mathematical theorem automatically. If given some information (or mathematical axiom), the system will output the result (or mathematical theorem). The knowledge of this system is rather limited, and the questions that can be solved are also limited. When facing more complicated practical problems, its limitations are exposed.

4.2 The second stage featuring the expert system

The second stage lasts from 1977 to 2006. In this stage, people are dedicated to importing more knowledge into computers, enabling them to solve more complicated practical problems. Artificial intelligence systems started to become more and more professional. The landmark event is the appearance of expert systems in various fields. The expert system is fed with masses of professional knowledge in a certain field, hoping it will solve some practical problems. Therefore, the development of artificial intelligence witnessed a new high tide, an eruption of expert systems in various fields. However, with the rapid increase of knowledge, it is getting increasingly harder to summarize a large amount of knowledge and feed it to the expert systems. In the meantime, the rate of conflict and contradiction in the knowledge base noticeably raises, causing a descend in usability. In all, the massive knowledge obtained makes the bottleneck of its development in this stage.

4.3 The third stage featuring the deep learning

The third stage lasts from 2007 till now. In this stage, thanks to the improvement of hardware and algorithm, the AI system's ability for acquiring and learning knowledge has greatly improved. The landmark event in this stage is the occurrence of deep learning algorithms based on big internet data. Using this algorithm, we have made a breakthrough in technologies such as calculation vision, voice recognition, and natural language processing. Thus, here comes a system that can automatically find knowledge, use it for self-training, and establish a self-decision procedure. Such systems are already put into practical use in various fields, pushing the development of AI to witness new high tides. For now, we are still at the early stage of this development tide. In the following 10 years, AI technology will achieve an appliance in a wider range. However, it is necessary for us to know that although the technology and appliance of AI have made some breakthroughs in the third stage, we are still in the "weak AI" stage, meaning it is not actually "intelligence" still lacking "independent consciousness" It can only solve certain problems under certain rules, so it is still far from "strong AI."

5. Three major schools of AI

The concept of AI was first put forward in 1956 at a Dartmouth meeting. The theoretical thought gradually developed into three schools: connectionism, behaviorism, and symbolism. All the three schools have already started in-depth study and applied their researches in such fields as photo recognition, natural language processing, and voice recognition. Every school has made great achievements in AI researches and had some shortcomings as well in the meantime.

From the ancient Greek people that put Euclid geometry together into the Euclid axiom system to Newton's mathematical principles of natural philosophy, human's modern mathematics and physics knowledge is organized into axiom systems. The main idea of symbolism is to use logical reasoning rules to inference the whole theoretical system basing on axiom. In 2011, Watson of IBM, an artificial intelligence expert system based on symbolism, beat mankind to win the championship in the television knowledge competition jeopardy. However, symbolism faces four main challenges: automatic acquisition of knowledge, the automatic union of multiple knowledges, knowledge-oriented representation learning, and knowledge reasoning and application. Although symbolism realizes artificial intelligence by simulating people's thinking process, it is difficult to make a breakthrough in these four problems. In 1959, Hubel and Wiesel proved that the central visual system has the function of complex patterns composed of simple patterns by observing the response of cat visual neurons. Later, people gradually found that the visual center is a ladder cascade with a hierarchic structure. The low-level region recognizes the local features at the pixel level in the image, and the low-level features are combined into global features to form a complex pattern in the high-level region. The abstraction of the pattern is gradually improved until the semantic level. The basic idea of connectionism is to simulate the neural network of the human brain and design the artificial neural network into a multilevel structure, making the low-level output to be used as the high-level input. However, this method is limited, for the reason that it gets good results only when it is in differentiable, strongly supervised learning and a closed static system, and the training results are also limited to the problems with given conditions. Behaviorism promotes the evolution of machines by constantly imitating the behavior of human or biologic individuals and surpassing the original performance. It mainly depends on the reinforcement learning method with reward and punishment control mechanisms. However, the disadvantage of this method is that it oversimplifies the process of human behavior, ignores the internal activity process of human psychology, and ignores the importance of consciousness.

6. Flaws of artificial intelligence

The advantage of artificial intelligence lies in the huge amount of information storage and high-speed processing speed, but it cannot deal with such problems as Hume's question: Can we deduce "should" from "being" or deduce "value" from "fact" proposition? Nor can it handle the expression of emotion. Artificial intelligence tries to express people's emotions and signification through big data and gradually upgraded algorithms, but there is still no way to make a leap.

AI applies a "returning to modify" mode. In other words, a group of codes are meant to solve a problem. It used to be in code execution that the program ends when the problem is not handled properly. For artificial intelligence, if the problem is not handled properly, the code will automatically return to modify the data code before execution again. Repeated modification, which is also known as repeated learning, is artificial intelligence. Of course, good artificial intelligence technology can modify insufficient models within a certain range and then simulate human-specific functions to a certain extent, such as partial human computing and logical reasoning ability, but it is powerless to human "nonfamily similarity" analogy and decision-making ability. Therefore, the "person" in artificial intelligence is not a real "person."

Artificial intelligence has its fatal defects that the rational logic is limited and cross-domain ability is weak. AI cannot understand the relationship of equality, especially the relationship of value equality in different facts. AI cannot understand the inclusion relationship, especially the value inclusion relationship in different facts (the small can be greater than the large, and nothing can be derived from the existence). People can use informal and incorrect methods and means to achieve formal and correct purposes, and they can also use formal and correct methods and means to achieve informal but correct intentions. People can use ordinary methods to deal with complex problems but also (deliberately) use complex methods to solve simple problems.

Historically, artificial intelligence can be divided into three main schools. First, connectionism, which imitates the neural networks and the connection mechanism and learning algorithm in cortical neural networks, is mainly manifested in the deep learning method. In other words, the processing structure of multiple hidden layers is used to process all kinds of big data. The second is the behaviorism that imitates the behavior function of human or biologic individuals and groups and the perception-action control system, which is mainly manifested in the reinforcement learning method with reward and punishment control mechanism; that is to say, the representation of output planning is realized through the feedback of behavior enhancement or weakening. The third is symbolism, which is represented by the assumption that the physical symbol system (i.e., symbol operating system) has sufficient and necessary conditions to produce intelligent behavior and the principle of bounded rationality, which is mainly manifested in the application system of a knowledge map; namely, it is to process all kinds of information and knowledge by simulating the logical structure of the brain. These three artificial intelligence factions learn from each other, combined with the Monte Carlo algorithm, which is one of the two random algorithms that makes it possible for artificial intelligence systems in specific fields to surpass human intelligence, such as IBM's Waston Q&A system and Google DeepMind's AlphaGo system. (Monte Carlo algorithm should be used if the problem requires limited sampling, and a solution that does not require to be the optimal one must be given. On the contrary, if the problem requires that the optimal solution must be given but there is no limit on sampling, it is necessary to use the Las Vegas algorithm). Although these artificial intelligence systems have achieved remarkable performance, there are still many defects and deficiencies, and there may exist great hidden dangers.

First, let us analyze the connectionism that makes artificial intelligence hot at present. The main source of strength of the current artificial intelligence is the deep learning method proposed by Hinton in 2006, which greatly improves the efficiency of image recognition and speech recognition, and it effectively plays a supporting role in some industries of unmanned driving and "wisdom+." However, any algorithm has its incompleteness, and a deep learning algorithm is no exception. This method is best used under the tasks of differentiable (continuous function), strongly supervised learning (good calibration of sample data, constant sample category/attribute/evaluation target), and closed static system (less interference, good robustness and uncomplicated), while for nondifferentiable, weakly supervised learning (large deviation of sample distribution, many new categories, serious attribute degradation and diverse objectives) or in the open, dynamic environment, the effect of this method is poor, and the computing convergence is not good. In addition, compared with other machine learning methods, the model generated by deep learning is very difficult to explain. These models may have many layers and thousands of nodes. It is impossible to explain each node separately. Data scientists evaluate deep learning models by measuring their prediction results, but the model architecture itself is a "black box" which may make you unknowingly lose the opportunity to "find mistakes" Moreover, current deep learning technology has another problem. It needs a lot of data as the basis of training, but the results of training are difficult to be applied to other problems. To properly solve these problems in various realistic situational tasks, we need to combine other methods to learn from each other and coordinate.

The second one is reinforcement learning in behaviorism, the main advantage of which is that it can accumulate learning performance according to the gains and losses in interaction, which is similar to the real human learning mechanism. The main drawback of this approach is that it takes a very simplistic view of the process of human behavior. In the experiment, we only measure the simple reward and punishment feedback process, and some conclusions cannot be transferred to real life, so the external validity is not high. In addition, behaviorism is keen to study observable behavior, but it does not study the internal structure and process of psychology, which limits the in-depth development of artificial intelligence, due to its extreme proposition, denying the importance of consciousness and opposing consciousness and behavior.

The last one is symbolism and its knowledge graphs, which belongs to the category of modern artificial intelligence. The intelligent simulation method based on logical reasoning simulates intelligent human behavior. This method is in essence to simulate the abstract logical thinking of the human brain. By studying the functional mechanism of the human cognitive system, using some symbols to describe the human cognitive process, and inputting this symbol into a calculation that can process symbols, we can simulate the human cognitive process to realize artificial intelligence. The idea of symbolism can be simply summed up as a concept of "cognition is calculation" From the perspective of symbolism, knowledge is a form of information and the basis of intelligence. Knowledge representation, knowledge reasoning, and knowledge application are the core of artificial intelligence. Knowledge can be expressed by symbols. Cognition is the processing process of symbols. The reasoning is the process of solving problems by using heuristic knowledge and heuristic search, and the reasoning process can be described in some formal language. Therefore, it is possible to establish the same theoretical system of knowledge-based human intelligence and machine intelligence. At present, the main challenges in the field of knowledge maps include automatic acquisition of knowledge automatic fusion of multisource knowledge, knowledge-oriented representation learning, knowledge reasoning and application. Symbolism advocates the logical method to establish a unified theoretical system of artificial intelligence, but it has encountered the obstacles of "common sense" and the problems of knowledge representation and problem-solving of uncertain things. Therefore, it has been criticized and denied by other schools.

From the analysis of the characteristics and shortcomings of the above three schools of artificial intelligence, it is not difficult to find that human thinking is difficult to be explained in the existing theoretical framework of artificial intelligence. How can we find a way to a bright future for intelligent scientific research? The following is the bottom thinking and discussion on this problem.

Wittgenstein, a friend and teacher of Turing, the father of artificial intelligence, wrote in the first sentence of his famous theory of logical philosophy, "The world is the sum of facts rather than the sum of things" in which facts refer to the relationship between things, and things refer to various attributes. From the current trend of development, most of artificial intelligence technology is focusing on the work of identifying the attributes of things, such as voice, image, location, speed, etc. There is little work involving various relationships between things, but people have started to do the work, such as big data mining. In this dazzling artificial intelligence technology, people often think about such a question: what is intelligence? What is the definition of intelligence?

As for the definition of intelligence, some people say it is nonexistence, some say it is complacence, some say it is flexibility, and some say it is robust adaptation. There may be 100 statements as long as there are 100 experts. In fact, it is impossible to form a definition acceptable to all. However, this does not affect the consensus on some difficulties and hotspots in intelligence research, for example, information representation, logical reasoning, and autonomous decision-making.

After the appearing of data and information, intelligent information processing architecture becomes particularly important. So far, many experts have put forward some classical theories or models. For example, in the field of vision, David Marr's three-tier structure is still pursued by many intelligent science and technology workers. As the founder of visual computing theory, David Marr believes that the information processing done by the nervous system is similar to that done by machines. Vision is a complex information-processing task with the purpose of grasping various situations of the external world that are useful to people and expressing them. This task must be understood at three different levels: computing theory, algorithm, and mechanism, as shown in Table 1.1.

Some basic concepts proposed earlier by David Marr have become an almost perfect theory at the level of computing theory. The theory features that it tries to make the research of human visual information processing more and more rigorous to make it a real science.

At present, in the theory of explaining the working mechanism of the human cognitive process, the ACT-R (Adaptive Control of Thought—Rational) model proposed by Professor John Robert Anderson of Carnegie Mellon University is considered to be a very promising theory. The theoretical model holds that the human cognitive process needs four different modules, namely, target module, visual module, action module, and descriptive knowledge module. Each module works independently and is coordinated by a central generation system. The core of ACT-R is the descriptive knowledge module and central generation system. The descriptive knowledge module stores the long-term knowledge

Table 1.1 Three-layer structure of David Marr's computing vision.

Computing theory	Algorithm	Mechanism
The definition of information processing problem: its solution is the goal of calculation. The abstract nature of this calculation is characterized. Finding out these properties in the visible world constitutes the constraints of this problem.	This includes research on the algorithm used to complete the expected calculation.	The physical entity that completes the algorithm consists of the given hardware system that forms the framework of machine hardware.

accumulated by individuals, including basic facts (such as the sentence of "Seattle is a city in the United States"), professional knowledge (such as "design method of high-speed railway traffic signal control scheme"), etc. The central generation system stores individual procedural knowledge, which is presented in the form of condition—action (production) rules. When certain conditions are met, the corresponding actions will be executed by the corresponding modules. The continuous triggering of production rules can ensure the cooperation of each module and simulate the continuous cognitive process made by individuals. ACT-R is a cognitive framework used to simulate and understand the theory of human cognition. ACT-R attempts to understand how humans organize knowledge and produce intelligent behavior. ACT-R aims to enable the system to perform various human cognitive tasks, such as capturing human perception, thoughts, and behaviors.

Just as many models that explain and simulate human cognitive process, both the three-tiered computing vision theory of David Marr and the ACT-R theoretical model proposed by John Robert Anderson have a common disadvantage: they cannot organically unify subjective parameters of human and objective parameters in machine/environment, and the model is not flexible. It is difficult to actively generate robust adaptability, not to mention higher-level representation and evolution of emotion and consciousness. Compared with human beings, there are inherent deficiencies in a more basic philosophical level in addition to the limitations of input representation and fusion processing. In other words, it cannot answer Hume's question.

Hume's question was put forward by the British philosopher David Hume in the first volume of *A Treatise of Human Nature*. At the beginning of the books, he puts forward a philosophical problem that has not been well solved, mainly referring to the problem of causality and induction, that is, whether the so-called "should" can be deduced from "be" whether the "fact" proposition can deduce the "value" proposition. Hume pointed out that the knowledge obtained by causal reasoning constitutes the vast majority of knowledge that human life depends on. This question raised by Hume's reflection on the universality and inevitability of causality was called "Hume's question" by Kant. Hume's problem is a famous philosophical problem on the surface, but it is actually a bottleneck and difficulty of artificial intelligence. When data is represented as information, the signifier is a relatively objective expression of being, and the signifier is a subjective expression of "should."

From the perspective of epistemology, "should" is to take the maximum or maximum value from the many values of the parameters (or variables) describing the state and characteristics of things, "be" is taking arbitrary value from the many values of the parameters (or variables) describing the state and characteristics of things. From the perspective of axiology, "should" is to take its maximum or maximum value from the many parameters (or variables) describing the value state and value characteristics of things, "be" is taking its arbitrary value from the many values of the parameters (or variables) describing the value state and value characteristics of things.

Due to the influence of preferences, habits, and customs, human epistemology and axiology often have noncausal induction and deduction. For example, strictly speaking, the value proposition of "a gentleman makes unremitting efforts to perfect himself" cannot be deduced from the fact that "heaven changes through movement" Still, as time goes by, this habit of analogy has gradually become something of causality. The advantage of artificial intelligence lies not only in its large storage capacity and fast computing speed but also in its unbiased mind and cognitive closure from the source. However, I am afraid it is still far from the perfect answer to Hume's question when it deals with those questions put forward by human beings. If artificial intelligence has certain intelligence, it should be more digital logic language intelligence. It can greatly improve work efficiency under the tasks of established rules and statistics in specific scenes and established output, but it is still difficult to create something out of nothing and make random changes in complex situations with emotion and intention. In the future, the development trend of intelligent science will be the continuous integration and promotion of human-machine intelligence.

CHAPTER 2

Human-machine hybrid intelligence: standing on the shoulders of intelligence

Contents

1. Human-machine hybrid intelligence

The theory of human-machine hybrid intelligence focuses on a new form of intelligence produced by the interaction of human, machine, and environments. It is not only different from human intelligence but also different from artificial intelligence. It is a new generation of intelligent scientific systems with a combination of physics and biology. Human-machine interaction technology mainly involves the physiologic and psychologic ergonomics below the human neck, while human-machine hybrid intelligence mainly focuses on the intelligent problem of the combination of the brain above the human neck and the "calculation" of the machine. Human-machine hybrid intelligence is different from human intelligence and artificial intelligence in the following three aspects. Firstly, at the intelligent input end, the idea of human-machine hybrid intelligence does not individually rely on the objective data collected by hardware sensors or the subjective information perceived by human facial features,

Integrated Human-Machine Intelligence
ISBN 978-0-323-99562-7
https://doi.org/10.1016/B978-0-323-99562-7.00002-4

but it effectively combines the two and forms a new input mode with the prior knowledge of contacts. Secondly, in the stage of information processing (an important stage of intelligence generation), a new way of understanding is constructed by integrating people's cognitive style with the computing power of the calculation. Finally, at the output end of intelligence, the value effect embodied in people's decision-making is added to the calculation's gradual iterative algorithm, and by matching them to each other, an organic and probabilistic optimization judgment could be formed. In the continuous adaptation of human-machine integration, people will consciously think about inertial common sense behavior, and machines will find the difference of value weight from people's decisions under different conditions. The understanding between humans and machines will change from one-way to two-way understanding, and human initiative will be mixed with machine passivity. People deal with the subjective information of "should" and other value orientations they are good at, while machines not only deal with the objective data of "be" and other regular probabilities they are good at, but also optimize their own algorithms from people's processing of "should" information, to produce the effect of "people - + machines" which is both greater than people and greater than machines.

The human-machine hybrid adopts a layered architecture. Human beings analyze and perceive the external environment to form intentional thinking through their acquired cognitive ability. Their cognitive process can be divided into several parts: memory layer, meaning layer, decision-making layer, perception and behavior layer. The machine perceives and analyzes the external environment by detecting data. Its cognitive process has a target layer, knowledge base, task planning layer, and perception and execution layer, forming formal thinking. The same architecture indicates that humans and machines can fuse at the same level, and causality can also occur at the same level or among different levels. Fig. 2.1 is the schematic diagram of human-machine hybrid intelligence.

Man-machine integration intelligence, in short, is to make full use of the advantages of man and machine to create a new form of intelligence.

British Prime Minister Winston Churchill once said, "You can see the long future as far as you can see the past" Therefore, it is necessary for us to take a look at the past of human-machine intelligence integration. Any new thing has its origin, and human-machine hybrid intelligence is no exception. Human-machine hybrid intelligence mainly originates from the two fields of human-machine hybrid interaction and intelligent science, whose origins are closely related to the University of Cambridge in Britain. In the

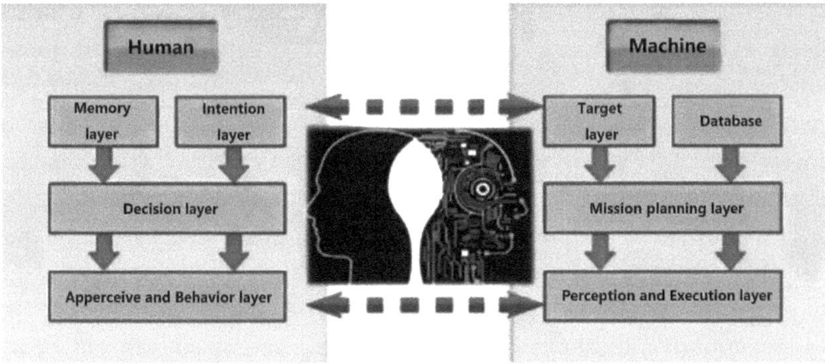

Figure 2.1 Human-machine hybrid intelligence.

summer of 1940, at the time that German bombers flew to London, the research prelude of human-machine hybrid interaction and intelligent science was slowly opened. To resist the German attack, the British began the scientific and technologic application of radar, aircraft, and password decoding. At that time, the first aircraft cockpit (the famous Cambridge Cockpit) to study human-machine hybrid interaction was established at St. John's College of Cambridge University to deal with the mistakes and faults made by pilots during flight missions. In addition, Turing, a graduate of King's College of Cambridge, led the work of decipherment of the German "Enigma" secret message. In fact, as early as the 19th century, Charles Babbage and Ada Lovelace from Cambridge University (daughter of the famous poet Byron, who graduated from Cambridge University and became the world's first programmer) began to cooperate in the development of mechanical calculation software and hardware. After the 20th century, Bertrand Russel and Ludwig Wittgenstein great contributions to the origin and development of intelligent science. Geoffrey Hinton, the father of deep learning in the field of human-machine hybrid intelligent integration, was a student in the Department of Psychology at Cambridge University; and the father of Alphago—Demis Hassabis—graduated from the Department of Computer Science at Cambridge.

In human-machine hybrid intelligent integration, it is especially important that humans should be able to understand how the machine sees the world and make effective decisions within the limits of the machine. On the contrary, the machine should also be "familiar" with the cooperating people. Just like the doubles teammates in some sports activities, it is impossible to produce appropriate integration and accurate coordination like chemical changes without tacit understanding between each other. Effective human-machine hybrid intelligence integration often means bringing people's thoughts to machines, which means that people will begin to consciously think about the tasks they usually perform unconsciously, while the machine will begin to process the partners' personalized habits and preferences. Both of them must also change with the change of the environment at anytime and anywhere; computers and psyche interact with each other just as bosom friends and give full play to their advantages and strengths. For example, human beings can break logic thinking patterns and use intuitive thinking to make decisions, and machines can detect signals that humans cannot detect with feelings. Every proposition understood by human beings must be completely composed of various components known to people.

The solution of intelligent mechanisms and mechanisms of human-machine hybrid will become the key to winning the war in the future. Any allocation will be limited by the scale and scope. The function allocation in human-machine hybrid intelligence is one part of such allocation, and the other part is the ability allocation. The function allocation is passive and caused by external demand, while capability allocation is active and internally driven. In complex, heterogeneous, unstructured, and nonlinear data/information/knowledge, human or human-like directional preprocessing is very important. When the problem domain is preliminarily reduced, the bounded, fast, and accurate advantages of the machine can be brought into play. In addition, when a large amount of data/information/knowledge is obtained, the machine can initially map them to several fields before people can further process and analyze them. The assimilational adaptation and reciprocal balance of these two processes are roughly the process of human-machine organic integration.

The key to the future work lies in the "being just to the good point" of the human-machine hybrid. It is impossible for machines to create either consciousness or the S-shaped separation line in the middle of yin and yang fish no matter how it makes efforts. No matter how many and big machines are, they are only half of the world relative to people. If you do not believe it, you might as well pay a little attention to the evolution of the current epidemic, and then you may get the point.

One evening, walking around the corner, the author heard the cuckoos of a cuckoo. It cannot help but remind people that people in different regions have different transliteration of the bird, such as "cuckoo, cuckoo" transliteration in the "grandpa and grandma, mowing wheat and planting grass" or "cuckoo, cuckoo" from "do not cry, it is hard to be single." In short, people can create all kinds of strange meanings at will, but for machines, only volume, tone, and timbre are three physical quantities that reflect the characteristics of sound.

2. Origin and future development direction of human-machine hybrid intelligence

The interaction among humans, machines, and environmental systems produces intelligence, which is not only a scientific problem but also includes the research of nonscientific parts (such as humanities, art, philosophy, and religion). Among them, the human is a complex system, but the machine is a relatively simple system, and the fluctuation of the environment is very large. Therefore, the human-machine environment system we study has both "certainty" and "randomness" which makes it a "complex giant system" Mr. Qian Xuesen believes that human beings have not found the general principles and methods to solve the "complex giant system" but the theory of a human-machine hybrid intelligent system may be a useful attempt. However, the inconsistency of space-time and logical mechanism in the human-machine process is the main difficulty of intelligent integration. The bottleneck of human-machine hybrid intelligence is how to realize reasonable, favorable, and rhythmic rhythm. For example, in the process of human-machine hybrid, some problems will become more and more prominent. For example, how does one allocate human-machine functions effectively? When, where, and how is the human-machine allocated? Is it better to take the human speed as the standard or the machine speed as the standard when the human and machine speeds do not match? How can human-machine hybrid intelligence integrate learning, understanding, decision-making, reasoning, perception, and intention? And how do data, information, knowledge, intelligence, and wisdom interact and transform? What is the relationship between these concepts? The *Book of Changes* analyzes movement, the *Tao Te Ching* discusses righteousness and wrongness, Sun Tzu's *The Art of War* talks about virtuality and reality, Wittgenstein proves existence and nonexistence, Boolean algebra says it as 0 and 1, Turing tests the similarity between human and machine, and von Neumann builds structures for storage and calculation. What is the relationship among these concepts? Is there a logical relationship other than "and" and "or" in digital logic? How does one define the calculation of analogy degree in nondigital logic? Is the idea of being "people-oriented" still right? If not, what is the orientation? If yes, will it split the common relationship among people, things (machines), and the environment? How does one represent the statements between "yes" and "no" between "should" and "should not" between "can" and "cannot" between "want" and "do not" between "positive" and "negative" and any other similar statements? How does one characterize macroscopic and mesoscopic superposition and entanglement states? How does one produce and cultivate tolerance, concession, compromise, and other mechanisms in a human-machine hybrid?

Contemporary artificial intelligence has developed from the original fully artificial compiled machine automation to artificial precompiled machine learning. The next development may be to realize machine cognition through the method of human-machine hybrid intelligence, and finally realize machine awakening.

3. Human-machine intelligence is neither human intelligence nor artificial intelligence

Once upon a time, a professor went to a remote country and went across the river by boat. He asked the boatman on the boat, "Did you learn some math?" "No." "Did you learn some physics?" "No." "Do you understand computers?" "No, I don't understand."

The professor lamented, "You do not know any of these three things, then you have lost half of your life." After a while, the dark clouds covered and the wind blew everywhere. The boatman asked, "Can you swim?" The professor said: "No." "Then you may lose your whole life."

Calculating the present and the future things according to past data is a common means of mathematics, and calculating the present and the past according to expectations of the future is the method of human intelligence.

We know much more than we say. What we do not know is much more than what we know. And we do not known what we do not know is more than we do not know.

Human sensory stimuli and the information they get are dynamically classified and clustered. They are not completed at one time, but have been gathered and changed many times (this reincarnation mechanism has not been clarified yet). The invisible Tao in the *Book of Changes* is fragmented and fluid. Therefore, it is the fragmented rules, probability, knowledge, data, and behavior that constitute human intelligence. In other words, it is generated and evolved in the strange situations of daily heterogeneous activities. Human intelligence is not formal and logical from the very beginning, and human logic is customized for nonlogical services. On the contrary, machines are organized and programmed from the beginning, and they also serve human nonlogical services.

In essence, the difference between the marking of data and the representation of information lies in whether there is meaning, that is, whether the possibility is understood. Although the representation system involved is formulated and endowed by man, it lost its original activity as soon as it was born, including various attributes and relationships, flexible connection, and combination under intentional participation. Many human representation methods often confuse God. For example, one flower has one world, and one tree hides one bodhi. In other words, each person has their understanding of the world. The flaw of knowledge map lies in the classification of knowledge. It rigidifies the original lively representation of knowledge, making it lose the flexibility of convergence of connotation and extension, just like the process of professional title evaluation. Expressing infinity with limitation is beauty, interpreting infinity with limitation is wisdom (truth), and connecting the two is goodness ("should" or "righteousness"). The decision-making of a machine usually uses appropriate dimensions to reduce classification information entropy. In real life, people's processing of information is the dimension of bivergentum. The length, degree of polymerization, quantity, and elasticity of the information all can change.

If the storage of a machine is a real configuration, the human memory is both a virtual and real configuration. And as time goes by, the virtual configuration becomes more and more, while the real configuration becomes less and less. Not only can it create nothing out of something, it can even create something out of nothing, like some legends or gossip in various history books. More interesting is that human memory can derive emotion, which is inconceivable for a machine to do.

Most of the human learning process is not only to get a clear answer, but it is more to find various possible ways to understand the world and discover the world. But the "purpose" of "learning" of the machine (supposing the learning has a purpose) is not to find the contact, but to seek a result.

The basis of intelligence is not the calculation but the law, is the method of understanding, and the way of understanding. Understanding is the key. Natural language processing will make no breakthrough if it only pursues the recognition rate without solving the understanding problems first.

Actually, people have the low recognition rate to the sound and often ask what other people said. Asking others what they have said is the most critical ability, because only when we know there is no understanding can we ask such a question. The understanding of many systems depends on people eventually. If no one involved, no understanding will appear no matter how many words have been dealt with. What the current AI misses is the bionic of the human perception, so we can not fully understand the physiologic and psychologic mechanism of human beings when they make decisions. All the aforementioned means that only when AI can feel the outside world like people, use the processor to think rationally like people, simulate and learn the human beings from the inside to the outside, can we say that such AI is perfect.

Game theorist Ariel Rubinstein published a collection of essays called "*Language and Economics*" In one of the papers, Rubinstein uses a game model to illustrate that the "debate" has great benefit for listeners who do not participate in the game because the "debate" forces both sides to disclose "private" information to the public. Mathematical methods may obscure deep insight, while people's intuitive perception, whose carrier is the sensory organ of the body, has contained the organism's understanding for various relations. To fix this understanding and form a "memory" people need the help of another ability, which is called "rational" ability. In the initial stages of the rational capacity, the "concept" begins to form. The concept is boundary, constraint, and condition, which will change a lot, and even go to the opposite in different situations. This is also why it is difficult to define intelligence, and all kinds of accidents happen in the activities in which human beings participate. Arthur Schopenhauer once pointed out that "Where the calculation begins, the understanding ends." This is because the calculator only focuses on the relationship between symbols fixed as the concept, no longer the changing causal process that happens in the real world. Opposed to pale "concept" thinking, Schopenhauer also gives wonderful discussion about the insight into "intuitive understanding" "Every simple person has reason. Just tell him what the premise of reasoning is and he will have reason. But understanding is different. It provides original things, and thus intuitive knowledge. There are natural differences between people. In fact, every major discovery and every historic world plan is the product of such a glorious moment. When the thinker is in the external and internal favorable environment, various complex and hidden causal sequences have been examined thousands of times, or unprecedented ideas have been blocked thousands of times.

Suddenly, they appear as the form of understanding." In this sense, all the current calculation intelligence, as long as it is not "sensory-based" intelligence, will never get creativity like human beings in the future. Here, the "senses" refers to the organ that makes a direct perception of the "world" the organ that has the ability to directly present the picture of the world, not like computers today: only with the help of human beings can they face the world to "reproduce" something. Qian Xuesen said, "The human body is like a system." First of all, it is an open system that has communication with the outside world. For example, material communication happens through respiration, diet, and excretion, while information communication happens through vision, hearing, taste, smell, and touch. In addition, the human body is composed of hundreds of millions of molecules, so it is not a small system or a large system, but a giant system larger than the large system. The components of this giant system are different, and the interaction between them is also extremely complex. So it is a "complex giant system." In fact, the current AI uses only a small part of the programmable human rationality, which is far from human rationality, not to mention the preliminary closer to the more magical part of human beings: sensibility.

Galileo once said, "Mathematics is the language describing the universe." More exactly, it should be that mathematics is one of the languages describing the universe. Besides mathematics, there are many other ways to describe reality. This is also a problem that intelligent science faces: how to effectively integrate grammar, semantics, and language use from the different languages. For multiple cognitive systems, common cognition ingredients are scarce and important, and mathematics is an attempt at this area to depict the correlation between objects (but not the only). If we talk about a different civilization, they use different ways to describe things, so the form will naturally be different. Mathematics is not the truth; it is just a statement of an aspect of reality, similar to the feeling of the blind when they touch elephant's legs. Mathematics and poetry are both the products of the imagination. To a pure mathematician, the material he faces is like a lace, a leaf of a tree, one piece of grassland, or a change of a man's face between the bright and dark light. In other words, the "inspiration" which has been called a "poet's fanaticism" by Plato, is as important to mathematicians. For example, when Goethe heard the news of Jerusalem's suicide, it was as a light flashed before him, and he immediately finished the outline of the book "*The Sorrows of Young Werther*" where he recalled, "This pamphlet seems to be written in the unconscious." And when the "Prince of Mathematics" Gauss solved one of the problems (Gauss and symbols) that had plagued him for years, he wrote to a friend, "Only a few days right before the last days, I succeed (I would say, I succeed not due to my hard exploration, but due to the God's favor), like the moment of lightning bombardment, this mystery was solved. I failed to know how the three things, my former knowledge, the way I last tried and the reason I succeed, are linked together." As another example, rewards and punishments are the core mechanism of machine reinforcement learning, while human learning has some other mechanisms besides rewards and punishments (adaptation, whether to actively reward/punish or passively reward/punish) as the stimulation; there are also transition processes such as choices between them. In addition, the human reward and punishment mechanism is much more complicated than the simplified machine reward and punishment mechanism, for there are not only reward mechanism, punishment mechanism, and even punishment-reward mechanism, but also some kind of punishment to express the real rewards (such as demote someone to promote power). Of course, a lot of situations are like a kick upstairs. Machines have not inherited many tricks from human beings at present except for binaries.

Walking through the endless traffic and retreating all over the body is a classic situation of man-machine situation coordination. When you think about it carefully, there is a similar relationship between situation and yin and yang: state is yang (explicit "being"), while trend is yin (implicit "should"). Sense (attribute) is yang, percept (relationship) is yin. There is yang in yin, and there is yin in yang.

The biggest difference between human learning and machine learning lies in common sense learning. When people educate or are educated, they are dealing with compound cognition, not just regular probabilistic input. People's common sense is very complex, involving physics, psychology, physiology, ethics, arts, and sciences and so on, including not only the topology of time and space, but also the topology of logic and nonlogic. Human beings are both dynamic and static. The machine is the same, but its dynamic and static statements are still different from those of humans. Man-machine hybrid learning, man-machine hybrid understanding, man-machine hybrid decision-making, man-machine hybrid reasoning, man-machine hybrid perception, man-machine hybrid intention, and man-machine hybrid intelligence are the future development trend and direction.

People have the ability to change variables into constants, reason into sensibility, logic into intuition, nonaxioms into axioms, individuality into commonalities, and confrontation into compromise. For example, people can not only express "how" with procedural knowledge, but they also express "why" with descriptive knowledge. As for those "what" "where" and "when" questions, the machine can help to assist in retrieval. Whether it is a human's natural intelligence or artificial intelligence, it finally involves the problem of value orientation. Unfortunately, machines will not have this problem in the visible range in the future. If price is scalar and value is vector, it can also be said that data is scalar, information is vector, machine is scalar, and man is vector. If data is scalar and information is vector, knowledge will be a vector of vectors; the reason is that data is physical and has no value in itself. Information is psychologic and has rich value.

At present, the reason why mainstream artificial intelligence theory loses its advantage is that the rational choice assumption that it is based on implies that the decision-making individual or group has homogeneity of behavior. This assumption ignores the differences among the things in the real world and the differences in people's understanding of the world under different conditions, which leads to a great reduction in the applicability of the mainstream theory, which is also why it can not include the "anomaly" in the scope of interpretation. To solve this fundamental problem, after years of research, many thinkers have gradually clarified the basic direction of deconstruction and reorganization of mainstream intelligent science, to bring the heterogeneity of individual behavior into the analytical framework of intelligent science, and to take the homogeneity of individual behavior as a special case of heterogeneous behavior under rational assumptions, to enhance its ability to explain and predict new problems and phenomena without losing the basic analysis paradigm of mainstream intelligent science. This means the heterogeneity of behavior is condensed into two basic assumptions. First, the individual is bounded rationality. Second, individuals are not only completely egoistic, but they also have a certain degree of altruism. Psychology, economics, neuroscience, social ecology, and philosophy provide a theoretical springboard and foundation for intelligent science to realize its heterogeneous behavior analysis. It can be simply called the "phenomenon of different people and the same machine" The future intelligence should grow up on the basis of a new generation of informatics integrating many disciplines, rather than just on the current mathematical basis that has much incompleteness.

Novices have no sense of abstract and boring information, while experts can extract dynamic, fresh, and different information from it. They are extremely sensitive, which means that they can see information that others cannot see (extract heterogeneity from homogeneity), perceive information that others cannot perceive, and form intuitive (fast) decision-making, which also leads to unusual irrational behaviors and beliefs. "Cognitive miser" means that the human brain likes to search for obvious surface information for processing when making decisions to save cognitive resources, to draw a conclusion quickly, but the result is likely to be wrong, so it is famous for its superficiality. Different from the "cognitive miser" there is also a concept in psychology called "fully disjunctive reasoning" which means that when you need to make a decision in the face of multiple options, or you want to draw the best solution according to hypothetical reasoning, you will analyze and evaluate the results of all options or possibilities to get the correct answer. Because of the systematic analysis, the speed is relatively slow.

Tacit knowledge has caused a lot of uncertainty, and the implicit rules make the interaction more complex. Its root lies in the fact that interactive objects have the characteristics of "they can constantly correct themselves in an uncertain and non-static environment" which requires not only the renewal of knowledge, but also the emphasis on tapping the potential of the organizational mechanism. In essence, human-machine hybrid interaction is the integration of human perceptual structure and human partial rational programming. "Compassion" is easily understood as that we share the feelings of others in this feeling in some way. In fact, empathy is a co-occurrence expectation of the consistency of emotional order. In the field of consciousness, we can find at least the following six different ways of "co-occurrence": co-occurrence of mapping, co-occurrence of empathy, co-occurrence of flow, co-occurrence of visualization, co-occurrence of symbolization, and co-occurrence of conceptualization. Therefore, although "co-occurrence" was first used by Husserl for other people's experience, it is actually a basic element running through all conscious experience structures. For this, machines are still far from learning to achieve it.

Hawking and Mlodinow described light as "behavior like both particles and waves" Intelligence is also a similar thing. Dispersion is like waves and aggregation is like particles. The object is static; the allocation matching is dynamic and constantly refreshed. It can be said that it changes every moment. How to grasp the changes of man-machine function analysis in different periods may be a very interesting problem. Now, many unmanned systems are not really unmanned, but there are no direct people, and the requirements for connecting people will be higher. The autonomy mechanism of human-machine hybrid integration in different situations is different, such as individual autonomy being different from system autonomy. In addition, an important problem of man-machine integration is how to balance, such as the balance of ability, opportunity, mode, and research and judgment. Poor integration is often caused by the imbalance of these aspects. For example, human-machine hybrid is divided into self and internal interaction and diplomatic interaction with others. Many expressions or representations only have logical meaning to other objects, which is often inconsistent with the psychologic meaning of the real sender. This situation often appears when there is lack of fluency, obscurity, and difficulty of human-machine hybrid deep-seated communication. In comparison, machines are good at dealing with family similarity, while people are better than dealing with nonfamily similarity, because human beings can extract acquaintance or similarity from unknown or similar things, and man-machine integration takes both into account. Discipline crossover aims to find nonfamily similarity for directed association. Wave particle duality is a continuous and discrete situation. Both situation and perception have duality, so is cognition. When it is discrete, it can cross borders and integrate nonfamily similarity. When is continuous, it often reflects parallel inertia to maintain family similarity. Whether the combination of human irrational cognition (discrete) and machine rational cognition (continuous) is consistent with justice (the thing that is right and should be done) is one of the main indicators to measure effective integration.

There are two difficulties in man-machine hybrid intelligence: understanding and reflection. Man is weak in state and strong in tendency, while machine is strong in state and weak in tendency; on the other hand, man is weak in feeling and strong in thinking, while machine is strong in feeling and weak in thinking. At present, the human-machine hybrid interaction does not have a high tacit understanding, because there is no one-way understanding mechanism, and the machine that has a humorous sense is still far away. In a table tennis competition, the athletes are able to calculate and do well, make their psychology not affect their technology (they are not afraid of losing if they want to win), and adjust their psychology (strength) to produce the best state; the psychological strength at the critical moment and the firmness of faith are all life characteristics that are difficult to produce by machines. In addition, the cooperation between man and machine must have combined expectation suanji, especially the appropriate second and third expectation suanji.[1] Self-confidence is trained through matching. The generation process of trust chain between man and machine is often as follows: strangeness → mistrust → weak trust → more trust → trust → strong trust → strongest trust. Without trust, expectations will not be generated, and without expectations, man-machine imbalance will occur. However, it is difficult to achieve integration through one expectation matching, so the second and the third expected compliance is likely to be the key issue of the consistency of man-machine hybrid. The premise of man-machine trust chain is that people should have self-confidence (this self-confidence is also trained by matching). After that, others' trust and trust in others mechanism can be generated. Trust in others and others' trust involve a multilevel expectation. If "being" is grammar, "should" is semantics, and the neutralization and addition of the two is pragmatics. Man-machine hybrid, which is the combination of grammar and semantics, discrete and continuity, clearness and roughness, self-organization and other-organization, self-learning and other-learning, self-adaptation and other-adaptation, autonomy and intelligence, is a compound of nonphysical cognition + embodied cognition community, calculation + method mixture, formal system + nonformal system. Response time and accuracy are important indicators of human-machine hybrid intelligence. Man-machine hybrid is machine-machine hybrid and machine mechanism + - brain mechanism; man-machine hybrid is also the man-man hybrid and human sentiment + human reason.

Artificial intelligence is relatively hard intelligence, human intelligence is relatively soft intelligence, and the integration of man-machine intelligence is soft and hard intelligence. Universal, strong, and super intelligence are all soft and hard intelligence, so human-machine hybrid intelligence is the future, but the hybrid mechanism is far from being clear. What is more, in a short time, not only people have evolved a lot, but the machine has changed too fast as well. The heterogeneity of individual and group behaviors is not only reflected in the fields of economics and psychology, but also becomes one of the most important issues in the field of intelligence. The current mainstream intelligence science is making a mistake made by economics before: it regards people as rational people. As everyone knows, people are living people, wisdom is living wisdom, and people have desires, motives, beliefs, emotions, and consciousness. But mathematical artificial intelligence is currently unable

[1] Suanji is a complex computing process of cross domain, multi-source and heterogeneous systems without the help of machines.

to do so. How to integrate these elements and transform it from a frozen and rigid state into a warm and flexible state should be the main criterion and scale to measure whether intelligence is smart or not. At the same time, it is also the current bottleneck and pain point that artificial intelligence is difficult to break out of, which is that they only have steel bars but not concrete. The integration of economics into psychology can turn a rational economic man into a perceptual economic man. However, it is not enough for the current intelligence science to be integrated into psychology. To be reasonable, intelligence also needs to be infiltrated into sociology, philosophy, humanities, art, and so on to realize the situation of evolving from the current state of rational intelligent people to naturally intelligent people. Intentionality in intelligence is produced by facts and values. When it is implicit, it is consciousness, and when it is explicit, it is called relationship. In this sense, the formalization of mathematics may kill intelligence. Wittgenstein believes that form is the possibility of structure. Objects are things that are stable and persistent, while configurations are things that change, not persistent. Wittgenstein also believes that we cannot deduce future events from current events. Superstition is precisely the belief in causality. In other words, there is no causal relationship between basic events or facts. Only those things that do not have any structure can be stable, immortal, and continue to exist forever, and anything that has structure must be unstable and can be destroyed, because they cannot exist when the components that make up them are no longer combined in the original way. In fact, there are two assumptions hidden behind every traditional choice (matching): program invariance and description invariance, which is also one of the reasons why the expected utility description is not deep enough. Procedural invariance indicates that preferences for prospects and behavior do not depend on the way these preferences are derived (such as preference reversal), while description invariance stipulates that preferences for selected things do not depend on the description of these selected things.

Recently, Chris Reid of the University of Sydney, Australia, did a research and believed: "They are redefining the nature of intelligence." A kind of yellow *Physarum polycephalum* called "Sponge Bob" can also remember, make decisions, predict changes, solve maze problems, simulate artificial transportation network design, and select the best food. They can do all these things, but they do not have a brain or a nervous system. This phenomenon makes scientists think it over again: what is the nature of intelligence? Through research, people have discovered that intelligence is the interaction phenomenon between the character's environment system, i.e., wisdom, wits, affection, intention, righteousness, change, heart, and so on. Psychologic heart is the interaction between human and machine environment system, and it is very difficult to perform psychologic restoration like physical restoration. The biggest difference between physiology/psychology and physics is that the former one is of a living thing, while the other is just a thing, the former one is of a thing alive and the other is of a thing that is not alive, and the former one is not easy to restore while the other is easier to restore. The reason why humanities and art are more likely to subvert original ideas than science and technology lies in the cross-domain reflexivity, which contains empathy and sympathy. It can surpass itself and be related to each other. People are generally unwilling to follow the old ways of life, so humanities art provides people with a broader way for imagination. It can be said that what people see is not important, and what matters is how people interpret what they see.

Derrida has a famous saying, "Give up all depth, and appearance is everything." It implies that life itself does not obey logic, because it is illogical and nonstandard, just like philology, dancing with an unfamiliar logic.

3.1 The secret in algorithm

1. The calculation in the algorithm includes two parts: calculation and suanji

(1) Calculation is logical, and logic is ratiocination.

① The ratiocination has rules.

② The rules generally do not change, but change is a kind of rule.

③ Rules are productive and belong to the category of automation.

④ The essence of automation is the logical reasoning of calculation, including AND, OR, NOT and their various combinations.

⑤ The physical basis of artificial intelligence is digital AND, OR, NOT and various combinations of calculation (but some nonlinear statistical probability computers are also involved).

⑥ Artificial intelligence is part of the field of automation.

(2) Suanji, which is nonlogical, is not ratiocination.

① Nonlogic things do not follow the reasoning procedure.

② Suanji penetrates various parts of inference fields.

③ This heuristic cross-domain ability is related to sensibility.

④ Sensibility is productive and belongs to the category of intelligence.

⑤ The core of intelligence is the nonlogical and irregular cross-domain perception of suanji, including the main AND, OR, NOT and various combinations and insights beyond.
⑥ Automation is part of the field of intelligence.

2. The method in the algorithm is the suanji of the suanji

(1) The method includes embodiment and reflexivity.
① Embodiment explains the cognitive process through concepts such as coupling and emergence, without assuming a concept of "representation."
② Reflexivity means that knowledge can produce knowledge, and behavior can produce behavior.
③ The blank-leaving in the painting and the blank-leaving in the talk are all methods, and the other parts are calculation.

(2) The method is not calculation.
① Method is not the rule of calculation, but the rule of suanji.
② The rule of calculation has its situation, but the rule of suanji has no situation.
③ Artificial intelligence has the property of closure, but intelligence has no closure property.

(3) The method of algorithm is more important than calculation.
① The method is not fact but value.
② Facts are suitable for reckoning, and value is suitable for perceiving.
③ The method can be got through counterfactual ratiocination, antivalue ratiocination, and cross-domain (non-)ratiocination.
④ Calculation is basic, bottom-up, and based on logic.
⑤ Method is complex, top-down, and based on understanding.
⑥ Method can help us see something far away, while calculation can help us see something close.
⑦ Man is good at method, while machine is good at calculation.

3. "Jisuanji" and deep situation awareness

(1) Calculation and suanji generate "jisuanji."
① Calculation uses "being" while suanji uses "should."
② Calculation has the basis, while suanji does not.
③ Calculation is science, while suanji is art.
④ The jisuanji is deep situation awareness.
⑤ Calculation is based on known conditions, while suanji is not fully based on known conditions.

(2) The deep situation awareness is insight.

① The situation contains calculation and suanji; the percept contains reflection and connection.

② Situation awareness is the use of certainty to calculate uncertainty.

③ The deep situation awareness is the jisuanji of facts, values, and responsibilities.

④ Calculation must be contextualized, scenario-oriented, and situational, while suanji can be noncontextualized, not scenario-oriented, and not situational.

⑤ In the process of jisuanji, an autonomous mechanism is derived, which is an appropriate switch between calculation and suanji.

⑥ The jisuanji can be changed, difficult, and simple, and it can also be assimilated, adapted, and balanced.

4. "Jisuanji" is not a scientific problem, but a problem of complexity.

"Machine" solves a large amount of problems of the same kind, and "Man" solves problems that are different kinds.

4. The use of man-machine hybrid intelligence

In 2018, the human-machine hybrid intelligence technology presented a trend of diffusion and aggregation of individual intelligence and group intelligence, which is related to both individual and group intelligence. Humans in man-machine hybrid intelligence are not limited to individuals, but also represent people-oriented cognitive thinking methods, including the public; machines are not limited to machine equipment, but they also include the mechanisms of computer systems. In addition, natural and social environments and real and virtual environments will all have an impact on the adaptability of man-machine hybrid intelligence. Human-machine hybrid intelligence focuses on solving the aforementioned details of the human-machine hybrid process.

In the case of "Avatar style of the integration of humans and robots" mentioned by Fast Company, the operator can see the scene captured by the robot with the help of a head-mounted display, and the robot transmits the feedback generated by the operator's actions back to the operator to form a closed loop of human-machine hybrid information. The remote-controlled robot transmits the robot's perception of the environmental situation, and humans handle understanding and decision-making. This is the initial stage of human-machine hybrid intelligence. The T-HR3 robot

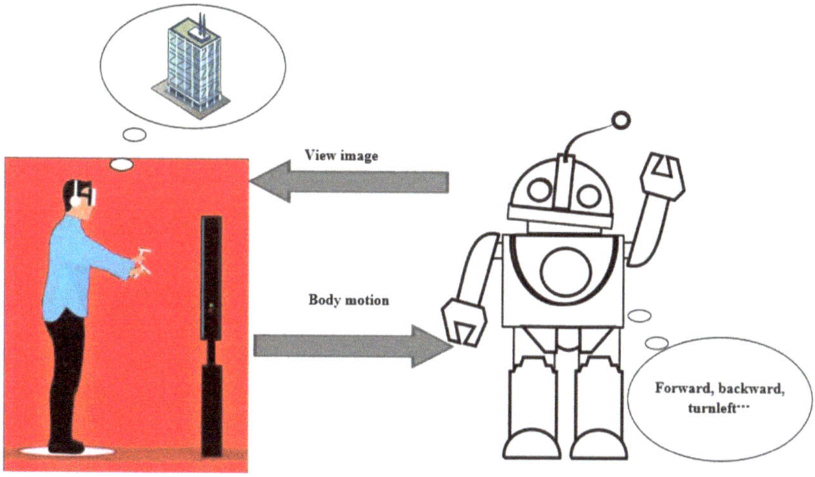

Figure 2.2 T-HR3 robot.

shown in Fig. 2.2 can use the latest 5G network technology to free the robot from delays in a long-distance working environment and can almost provide users with instant feedback.

This robot is also very accurate in conveying strength in the real scene. It can perform tasks that require strength to complete: holding the ball in both hands, picking up modules and stacking them, and even shaking hands with people. Under the practical application of man-machine hybrid and robots, the latest human-machine hybrid platform developed by Toyota will explore the safety management of the physical interaction between the robot and the surrounding environment, as well as a new remote control system that can map users' actions to the robot, which makes the integration of man and machine to obtain a better experience.

At the same time, man-machine hybrid intelligence in the manufacturing industry has also received attention and development. In the former industrial assembly line, robots replaced humans to complete repetitive mechanical work. The human-machine hybrid intelligence that appears in the manufacturing industry today is based on different hardware devices and environmental conditions. Some are similar to robot assistants, and some are exoskeleton suits. There is a "Miss Charlotte" man-machine hybrid robot in BMW's Spartanburg company that assists in the precise installation of the doors. Mercedes-Benz is also developing man-machine hybrid technology. The company customizes a more personalized service for each individual customer for luxury models, and it uses the combination of data and labor to make this service feasible. After the use of human-machine hybrid intelligence to replace the larger automation system, the special parts required by the customized version of the S-class car will not have the trouble of providing timeliness in the ordinary assembly line. Instead, they replace it with more convenient operation and management. MIT professor Julie Xiao is developing a special software algorithm whose purpose is to make robots understand the signals sent by humans, and then solve the communication problems between robots and humans.

In 1997, after "Deep Blue" won the chess battle between humans and machines, the US Department of Defense Advanced Research Projects Agency used it as a template and developed the next-generation robot with combat command and decision support system: "Deep Green" ("DG" for short). By adopting different combat plans through multiple calculation simulations, the "observation-judgment" link in the "observation-judgment-decision-action" loop demonstrates the possible effects of different battle plans, predicts the enemy's actions, helps the commander make correct decisions, shortens the time used to formulate and analyze combat plans, and takes the initiative to deal with the enemy instead of passively responding to an attack, so the US military commander can be one step ahead of potential opponents both in thought and action, as shown in Fig. 2.3.

The "Deep Green" system is mainly composed of a human-machine hybrid module named "Commander Assistant" a simulation module named "Blitzkrieg" and a decision-generating module named "Crystal Ball" Its architecture is shown in Fig. 2.4.

4.1 "Commander Assistant" module

The "Commander Assistant" module mainly deals with the function of the man—machine dialogue, which can automatically convert the commander's hand-drawn sketches and corresponding language with command intentions into brigade-level action plans (COA), helping to quickly generate combat plans and make quick decisions. This module includes the following three submodules: "Plan Sketch" "Decision Sketch" and "Automatic Scheme Generation."

4.1.1 "Plan Sketch" submodule

The "Plan Sketch" submodule has the following functions. It receives a user's hand-drawn sketches and voice input and converts them into standard military symbols, such as the US military combat symbol code MILSTD2525B. Commanders can think and draw in their own way,

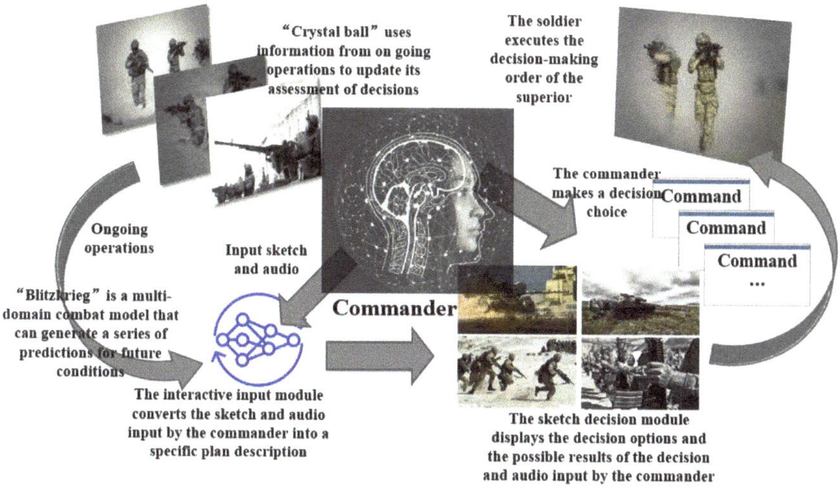

Figure 2.3 Principle of "Deep Green" system.

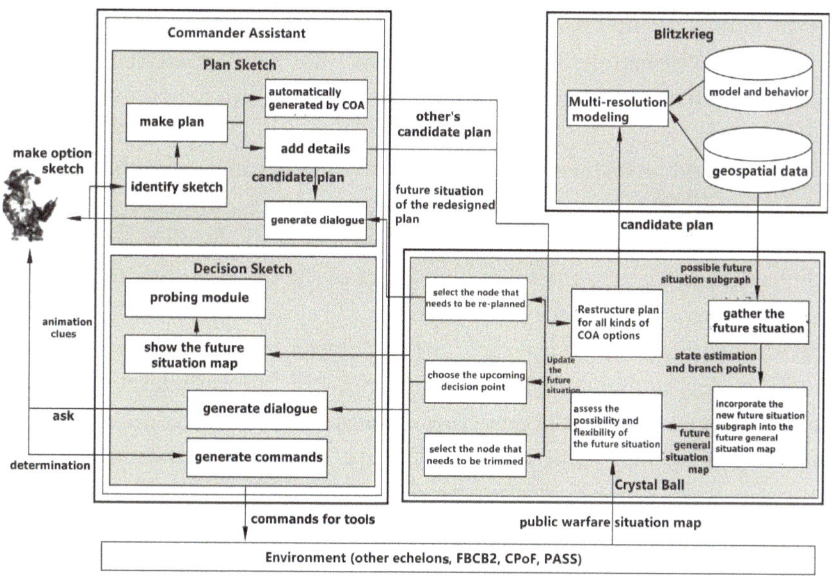

Figure 2.4 "Deep Green" system architecture.

instead of sticking to a completely formal military standard. It can also add details to the combat plan and have enough knowledge in various fields, and it can ask the user some questions to understand real intent and initialize the combat model when encountering a few unclear questions.

The output of the "Plan Sketch" submodule will be the action plan described in the military setting markup language. The "Plan Sketch" submodule includes sketch recognizer, plan inducer, plan automatic generator, detail adding planner, and dialogue generator. The sketch recognizer converts a series of self-drawn marks and voices into a series of standard military symbols; the plan inducer uses a large number of symbols to help the commander make plans and intentions; the detail adding planner will add details to the plan generated by the commander, so the "Blitzkrieg" module can simulate the program; the dialogue generator can interact with the commander to clarify ambiguities and help understand the commander's determination and intentions.

4.1.2 "Decision Sketch" submodule

"Decision Sketch" submodule is critical to achieving the "Deep Green" goal. Its purpose is to enable the commander to "see the future" It has the following functions: receiving input from the decision point of the "Crystal " and decisions from the commander, displaying the possibility, risk, value, effect, and other results of different decision-making options and multidimensional information, helping the commander better understand the situation that may be formed in the future, and conveying decision-making to subordinates.

Its submodule "Decision " includes the exploration module, the presentation module, the dialogue generator, and the command generator. The exploration module allows the commander to explore a possible future combat image to grasp the follow-up effect of the decision; the presentation module converts the information from the future combat image into an intuitive representation; the dialogue generator presents the required decision to the commander and communicates with the commander until the commander's combat intention is truly understood; the command generator expresses the commander's decision-making specifications as instructions to his subordinates and provides this information to the "Crystal " module to maintain and update future combat images policy.

4.1.3 "Automatic Scheme Generation" submodule

In the early days of the "Deep Green" program, the "Automatic Plan Generation" submodule simply transformed the commander's intentions into combat plans. With the advancement of the "Deep Green" plan, the goal of this module becomes to creatively and automatically generate a combat plan that meets the commander's intentions.

4.2 "Blitzkrieg" module

The "Blitzkrieg" module is the simulation part of the "Deep Green" plan. By using qualitative and quantitative analysis tools, various decision-making plans proposed by the commander can be quickly simulated to generate a series of possible future results. This module has a self-learning function, and the ability to predict future results can be continuously improved.

The "Blitzkrieg" module can identify each decision branch to predict the range and possibility of possible results, and then simulate each decision path. The "Blitzkrieg" module mainly includes three parts: a multidecision simulator, a model and behavior database, and a geo-spatial database. It has the following functions: inputting the plans of all parties in the war, determining the decision branch point or possible future situation, reasoning and evaluating the decision-making branch, and traversing all possible decision choices and continuously simulating all decisions.

4.3 "Crystal Ball" module

The "Crystal Ball" module will make timely and more accurate predictions of the future combat process based on the information during the combat process. Its main functions are as follows: receiving the decision-making plan from the "Planning Sketch" submodule in the process of generating possible future results, sending it to the "Blitzkrieg" module for simulation, and receiving feedback from the "Blitzkrieg" module quantitatively, conducting a comprehensive analysis of all possible future results in a quantitative form; obtaining updated information from ongoing combat operations and updating the possible parameters of various possible future results at the same time; providing the commander with the most likely future results with these updated possible parameters to analyze and compare possible future results; and reminding the commander to make a decision by using the analysis results to determine the upcoming decision point and calling the "Decision Sketch" submodule.

CHAPTER 3

The essence of intelligence is not data, algorithms, computing power, or knowledge

Contents

1. The essence of intelligence

The interaction of physiology realizes life, the interaction of psychology makes oneself, and the interaction of the character (machine) environment system derives the self in society. The interaction produces the real and the virtual. Interaction forms the "I," and "I" means interaction. There is no data, information, knowledge, reasoning, judgment, decision-making, situation, nor perception without interaction. First, the interaction process is two-way, that is, A gives B, and B also gives A. Second, the interaction process is proactive, and there is an equal initiation relationship between A and B. Third, the interaction process has the same rationality, and A should consider B at the same time, while B must consider the tolerance of A. Finally, the interaction process is purposeful, and there is a consistent and coordinated relationship between A and B. Therefore, in a strict sense, the current machine itself is not interactive: the machine does not have the abstract concept of "I."

It can also be said that intelligence is the existence that arises from the interaction, "I." Intelligence has little to do with data, information, knowledge, algorithms, and computing power. However, it has a lot to do with the interactive ability to form data, information, knowledge, and process and understand it. Data, algorithms, computing power, and

Integrated Human-Machine Intelligence
ISBN 978-0-323-99562-7
https://doi.org/10.1016/B978-0-323-99562-7.00003-6

knowledge are only part of the intelligence, so it is bit like moving a ladder to the moon with such things to achieve intelligence. Intelligence is closely related to the representation of nonexistence, the hybrid of belief and understanding, and the decision-making of facts and values. Intelligence is a variable interaction that can remove subjectivity. It can synthesize different existences, situations, and tasks, and realize the random switching from being rigid (like carving on the boat to seek the sword falling into the river) to acting by chance, from being unilateral (like blind persons touching an elephant) to being comprehensive, from "Cao Chong's proclaiming an elephant" to "blessing in disguise" so a trustworthy and interpretable primary form of intelligence (such as artificial intelligence) is gradually transformed into the field of predictable and adaptable human-machine environment system hybrid intelligence.

The key to the question why interaction is the source of intelligence lies in two aspects: "interconnection" and "mutuality" The so-called interconnection refers more to factual rounds, including physical, psychologic, ethical, mathematical, physical, and management, while the so-called "mutuality" is a more value-oriented concept, which includes not only initiative, intentional, and purposeful but also two-way, empathetic, and sympathetic aspects. Nonexistent "being" is a problem of contingency. The role of intelligence comparison is to find out and fill in the gaps in the interaction and to "judge Hercules from his foot" Belief and understanding are a kind of cognitive problem, and the role of intelligence comparison is to balance the contradiction between preconceived and gradual progress in the interaction. Facts and values are a practical problem, and the role of intelligence is to carry out the timely dialectic and accurate implementation of objective existence and subjective consciousness in interaction. Finally, the consistency of experience and experiment, prior and posterior, experience and inspection, and experience and nonexperience can be achieved through the "interconnection" and "mutuality" between the human-machine environmental system.

If "interconnection" corresponds to an actual number, and "mutuality" corresponds to an imaginary number, "interaction" corresponds to a plural number; if "interconnection" corresponds to facts, and "mutuality"-corresponds to value, "interaction" corresponds to intelligence (wisdom). It includes not only factual logical computers, but also value-intuitive (nonlogical) computers. Just as von Neumann summarized the steps of proof theory defined by Hilbert, "meaningful formulas" do not mean being true; $1 + 1 = 1$ is as meaningful as $1 + 1 = 2$, because whether a formula is meaningful or not has nothing to do with that one of them is true and the other is false. In this way, the intelligence generated by "interaction" is a set of formal mathematical multiple symbol systems and a set of intentional heterogeneous nonsymbol systems of human nature. These two systems will be established in a complex system of knowledge, fun, and flexible intelligence based on negation, equality, and implication.

Simply put, the machine (intelligence) is the acceleration of human-specific (rational) intelligence. The best machine is also related to people who use it. The combination of different humans and machines produces different effects. The human-machine hybrid can double the efficiency of the machine, or reduce the role of the machine, and vice versa. The primary function of a human-machine hybrid can solve various problems of consistency of change human beings are metaphysics and machines are The human-machine hybrid expands not only human senses such as vision, hearing, touch, smell, taste, etc., but it also enhances perceptual behaviors such as understanding, learning, judgment, decision-making, compliance, and assimilation. And more importantly, it produces a new form of intelligence, a new way of looking at the world: cognition + calculation.

Intelligence may be the process of resolving the contradiction between cognition/suanji and supply, and calculation from contradiction to accuracy. The calculation in cognition is human rationality, and this is the connection between machines and humans. How to add a little cognition in calculation is a key problem, and it is also the reason why it is difficult for machines and humans to communicate with each other. The cognition in calculation can be simplified to a question: how do we make the machine produce computing intuition? People produce flexible understanding based on intuition, and further refinement is the cognition in the calculation. Following this path, we could make the problem an entry point: how to improve the machine's multiview understanding and the machine's understanding of multiview crossing. The process philosopher Whitehead discusses understanding from the perspective of creativity, and the cognitive scientist Hou Shida has also studied the concept of flow. The possibility of a concept itself is the abstract process of type-analogy-analogy. The meaning of a concept is multiangle and multidomain. It is unrealistic to make only one standard with fixed representation. Just as the Chinese poem goes, "We will view the mountain as ridges and peaks with different distances and heights." Specific analysis and abstraction of specific problems are the essential characteristics of human intelligent cognition. The inexplicability of the complex algorithms of artificial intelligence is firstly due to the dynamic changes of knowledge and concepts. Artificial intelligence can generate machine intelligence differently from what humans do, but the field of human-machine integration is indeed the touchstone for artificial intelligence to advance to the advanced stage. DNA is a double-helix interactive structure, while intelligence is the multihelix interactive structure of system of the human, machine (machine is an artificial), and environment. Human-machine hybrid intelligent technology transforms people as well as things and the environment, which is a subjective and objective parallel technology.

The Taoist concept of "Tai Chi" is incomprehensible to Westerners. In the hearts of Westerners, one is one and two is two. What is meant by "one can be two, and two can be one" Big is big, small is small, what is meant by "both big and small" How can a definition contain two opposite things? "Nonduality" is a Buddhist term and a Chinese vocabulary that means no difference (from the "*Dictionary of Buddhism*" One real principle, such as equality, without being different from each other, is said to be non-difference). "Wisdom is the mind of right and wrong" is a Confucian concept, which is often combined with "benevolence, righteousness, etiquette, and faith" emphasizing that intelligence is not only cumulative knowledge but also cross-cutting knowledge. According to *"The Art of War"* by Sun Tzu, only with a "contingency" mindset can we see the essential laws. Do not use the appearance to deny the essence. Sometimes the appearance extends the essence, but more often, it will "cover" the essence. In some cases, appearance does not represent the essence or is even contrary to the essence. If there is no flexible thinking pattern, you will be bound by the wrong angle and factors and make the wrong decision. The thoughts of the Eastern people are often far from traditional Western rationalism, such as the nonwrong or nonambiguity in mathematics and the absolutism in classical physics. These rational thoughts do not involve contradiction and contradiction. There are often many discrepancies with objective practice. Some people think that mathematics cannot give general intelligence, and mathematics itself is a product of general intelligence. So can one person generate another person without interacting? Can one thing produce another thing without external force? Will a piece of knowledge mutate without being practised? Can a piece of data appear without being collected? Does one formula derive another formula without calculation? But, it is embarrassing that mathematics requires no contradiction (no ambiguity) from the beginning. During the French Enlightenment period, Voltaire, a famous philosopher and writer, once said, "Uncertainty makes people uncomfortable, and certainty is absurd." The same person can be given various identities in different situations and environments, for example, elder sisters, eldest sisters, aunts, and wives, even men disguised by women. From Zhuangzi's point of view, various things have their own two sides, and these two opposite sides coexist and depend on each other. Therefore, "the saint does not follow the path of dividing right from wrong but observes and compares the nature of things (conforms to the development of things themselves)." This shows that the debates between Confucian and Mohist on the "right" and "wrong" failed to see the essence of the development of things and took the wrong path and became farther and farther away from the essence. Zhuangzi believes that there is no opposite side to each other. The essence is the hub of the avenue. Grasping the hub of the avenue is grasping the key points of things, thus adapting to the constant changes of things. This point of view of Zhuangzi equals Lao Tzu's "Shouzhong" expounded in the "*Tao De Jing*" which means that things will naturally develop following the law if the key point in the opposition of things is found and then carefully maintained.

The current human-machine hybrid intelligence is a process in which people put part of the clear intelligence into the machine and then add their unclear intelligence in it while completing the goal according to the changes in the external task environment. In the future, human-machine hybrid intelligence may also be coupled with the intelligence generated by the machine itself. The key to human intelligence and wisdom lies in change, communication and communication change representation, change goal, change reasoning, change premise, change the decision, and change action. In contrast, the change of the machine appears to be more rigid and patterned, and the relationship between change and communication is not handled well. "Can" is not only an ethical issue, but it is also a question of wisdom or a question that combines responsibility and intelligence. Therefore, the flexibility of natural human intelligence is embodied in the choice between "Tao" and "De" (virtue). It is the standard representation and embodiment of fact and value, and it is the gathering of "being" "should" "can" "want" "change" Some intelligence approaches gather and converge the expert knowledge base through deep learning on a neural network that represents the existing prior knowledge. However, they are not able to handle newly generated data and information. In other words, the posterior knowledge cannot be upgraded to a priori, and the hidden knowledge cannot be discovered. Therefore, its function is to integrate outstanding achievements but not to innovate. This is a bit like education. The task of a school is to teach knowledge to students (a bit like machine learning), but education is not just to teach knowledge. Education should unearth the logic behind knowledge, or something more profound. For example, when we teach computers, we actually have to think about what is behind the computers. We should first cultivate students' sense of numbers, and then teach them the concept of calculation, what is addition and subtraction, how to apply the knowledge, and finally help them form insight.

The essence of the human-machine hybrid is to deal with the relationship between "change" and "invariance" Chinese people often call it the "change" and "not change" in "the *Book of Changes*" The change and invariance of human beings are driven by value, while those of machines are often driven by facts. Although machines may also have some ideas and habits of the maker, they still cannot make meaningful choices and decision-making in the context of change. User portrait (user information labeling) is such an example, which is to abstract the complete picture of the user's information by collecting data in various dimensions such as users' social attributes, consumption habits, and preference characteristics and then characterizing the characteristics of users or products, analyzing and counting these characteristics, and tapping potential value information. But it is more critical to achieve dynamic user portraits.

The in-depth situation awareness in human-machine hybrid intelligence is not a problem of set in the mathematical sense, because the elements are nonisomorphic and nonhomogeneous, and there will appear similar (nondifferential) elements. So people can call it the universal set/pseudo-set problem. The research of modern in-depth situation awareness has shifted from the study of "state" to "situation" from simple "calculation" research to complex "jisuanji" research, and from objective research of "facts" to the "value" research (research on the situation awareness of human-machine hybrid). All of them have the same shift from an objective thing to a subjective thing: just as a saying goes, "an egg breaking from the outside to the inside is an omelette, and it may give birth to a life when it breaks from the inside." The expansion of an intelligent system is a product under the dual guidance of the needs of the objective world and internal logic. Just as we know, it is easy to code facts (space-time coding), but coding value is difficult. Information is valuable data, and it is a product of the interaction of humans, environments, and systems. How the four cycles in the situation awareness—state, situation, sense, and percept—generate resonance will be the key optimizing items in the OODA (observe–

orient—decide—act) loop. The state and the sense belong to the outer circle, while situation and percept belong to the inner circle. The mutual promotion of these two cycles is very essential. The outer circle must "see and hear clearly" while the inner circle means "to understand and judge accurately" Traditional topology mainly studies the invariant properties of geometric shapes under continuous deformation. The topology of cognition is studying the invariable nature of perception under the continuously changeable situation, including facts (two dimensions of time and space) and value types (three dimensions of emotion, intention, and responsibility).

At present, the bottleneck of intelligence remains to be deep situation awareness in the human-machine hybrid. For instance, the warfare in the future turns out to be more than intelligent but wise warfare, which is bound to break not only the formal mathematical calculation, but also the logical calculation of traditional thinking mode. It is a new-style calculation that combines with the complementation of human-machine environment systems: suanji system. The intelligent war is based on the online information and knowledge system of human, objects, and environment, where people fight with intelligent equipment and relevant fighting methods in the field of land, ocean, air, sky, electricity, network, cognition, and society. Generally speaking, it is a war supported by the technique of complementation of human-machine environment and intelligent cognition system. Intelligent cognition refers to a comprehensive optimizing process of sense, analysis, judgment, and decision with the lack of data, information, and knowledge or the condition of exclusion and disturbance. It involves many procedures like input, processing, output, and feedback. Intelligent perception is only the inputting stage of intelligent cognition. In a narrow sense, intelligent cognition refers to machine input, processing, output, feedback, and other processes, which is a formal symbol system without referring to objects (like math). That is the root cause why machine intelligence is incomprehensible, unable to learn through out life, and difficult to form common sense. In a broad sense, intelligent cognition refers to input, processing, output, feedback, and other processes of a human-machine environment system, which is a system that combines symbol systems with referent (like human natural language), formal symbol systems without referent (like math), and changes of systems and task environments. That is the root reason why generalized intelligent cognition (human-machine hybrid intelligence) is comprehensible, conscious, easy to cross domains, flexible, changeable, and beyond human intelligence.

The development of theories related to intelligent cognition has mainly experienced three stages. In the first stage, it is based on game theory/ operations research, cybernetics, information theory, system theory, and other related theories, and the main goal is to achieve auxiliary calculation. In the second stage, it is based on expert systems and intelligent optimization and other related theories, the main goal of which is to achieve auxiliary decision-making and reduce people's physical and psychologic load. In the third stage, it is based on machine learning (including deep learning, reinforcement learning, transfer learning, etc.), data mining, knowledge graphs, brain-like computing, and other theoretical achievements in the field of artificial intelligence, the main goal of which is to achieve autonomy and intelligence in decision-making. Due to the particularity of game confrontation, traditional intelligent cognition will be gradually shifted to the intelligent cognition of human-machine hybrid integration to achieve insight purposes such as hiding the truth and revealing the false, removing the false and keeping the truth, which is embodied in the two types of seven-dimensional human-machine hybrid in-depth situation awareness—facts (including three dimensions of space - + one dimension of time), and values (one dimension of consciousness + one dimension of emotion + one dimension of responsibility)—achieving the goal of "goodness" on the basis of "speediness" and "precision".

The highest level of cognition is to go beyond the feeling, including not only feeling for you, but you for the feelings as well. It is just like the thing that changes, granting, stimulating, and awakening are real-time while listening to good music, appreciating good photography, or accusing a great game and confrontation, but not like the thing that the programmers estimate, formulate, program, and draw based on rules, regulations, conditions, and premises. Of course, a fixed procedural part must exist in the whole process, yet it is the invariance in changes. The main content and task of human-machine-hybrid intelligent cognition research is to deal with these changes and invariance. Among them, the recursive relationship is very important. It means that the entity establishes a relationship with itself, that is, calls itself during operation. The recursion of machines is standardized, but the recursion of humans is nonstandard, and it is even more flexible, which enables people to generate sarcasm, subject and act in half-genuine and half-sham.

The dimensions of cognition can be represented by four dimensions: state, trend, sense, and percept. The state includes the number of time and space, the situation refers to rate of change, the sense is about initiative (expectation, effort), and the percept refers to value of knowledge. It is impossible that the same property belongs to and does not belong to the same thing in mathematics. This is the most certain of all principles. Therefore, those who make the argument regard this as a final opinion, because it is the source of all other axioms by its nature. In fact, people can see things either right or wrong, but machines cannot do so, so the hybrid of man and machine is necessary. In the world, there are connections without attributes of things. There are also attributes of things that are not connected. There are values with facts, and there are values without facts. Therefore, the study of deep situation awareness (DSA) should be focused on the deformation of the state, the variation of the trend, the change of the sense, and changes in perception.

Philosophy pays more attention to the raising of problems, while mathematics pays more attention to the solution of problems. The reason why it is difficult for DSA to calculate may be that it is inconsistent with the law of excluded middle in Boolean algebra. State is related to time and space, and situation is not closely related to time and space. Sense is related to facts, while percept is closely related to value. In fact, in the common sense of life, many things can belong to and not belong to a certain category at the same time. For example, a person can belong to and not belong to the category of "parents" at the same time. He/she can be a parent for their children, and he/she is not a parent for his/her wife/husband; a teacup can belong to and not belong to the category of "tools" They can belong to tools while used for drinking tea, but they are not tools when used as works of art. The bridge linking state, trend and sense, percept not only includes temporal and spatial changes, but it also involves equivalence, implication, and transformation between factual values.

What connects the state and the situation is the change, and the connection between the sense and the percept is also the change. There are both simple variables and compound variables, as well as system variables and the hybrid of the three human-machine variables, including only the complex state and trend of "the real state and the virtual situation + the virtual state and the real situation," but also the complex sense and percept of "the real sense and the virtual percept + the virtual sense and the real percept" In addition, it includes the complex situation awareness of "the real/virtual state and trend + virtual/real sense and percept" The essence of art is individualized, and the essence of intelligence is also individualized. At this point, the two are interlinked. The difference is that intelligence also has common laws that provide mathematics with a logical stage based on the axioms of conventions. The fundamental reason for the incomprehensibility of artificial intelligence is that mathematics is a symbolic system without referents. By the way, comprehensibility is the peculiar nature of the symbolic system of referents in human natural language. There is no one-to-one mapping relationship between symbols and objects and their properties. If the real-time nature of representation cannot be achieved, symbolism will not be able to be improved; if it cannot solve the problems of dynamic representation and nonaxiom logic, it will be difficult for AI to fundamentally resolve the problems of credibility and interpretability. Many related automation/intelligence systems are the comprehensive balance adjustment of several key parameters, and they often gain one side and lose the other side, but many people only talk about the gain and ignore the loss. The key to intelligence is how to transform uncertain and uncontrollable factors into deterministic and controllable factors.

Human-machine hybrid intelligence is the parallelism of humans and machines, where they exit in each other. Human-machine mixing is human-machine serial, where people stop when machines move and vice versa. Strictly speaking, assisted decision-making or assisted driving is the integration of human and machine. In this case, human and machine are working at the same time, and the good coordination between the two parties lies in the consistent backup state at any time. If one party cannot keep up with the other, the high accident risk of cold start and long delay may appear. Typing or weighing is basically a human-machine mixing, for man and machine move in series. Turing believes that the behavior of the calculator at any moment is determined by the symbols he observes and his "state of mind" at that time. Some people now are starting to realize the difference between human-machine interaction, human-machine mixing, and human-machine hybrid, so they may be able to have a more updated understanding of the mechanism of calculation, perception, cognition, and insight if they go deeper.

News needs to be linked to each other to be clearly understood. What I remember most clearly is what Mr. Ruan Cishan said, "Things that seem to be irrelevant are actually connected internally." The same goes for intelligence and antiintelligence. Turing machine is essentially a finite automata, while humans are an infinite selection machine. The current human-machine hybrid intelligence should be the effective collaboration of the finite automata and the infinite selection machine. If the function reflects the relationship between states, then the trend is a matrix with members of functions, that is, the relationship of the relationship; if supervised-learning is learning of the state, unsupervised-learning is learning of the situation, and the situation is perceived through the situation. The DSA in human-machine hybrid intelligence may be a breakthrough.

Sima Qian said, "There are extraordinary people in the world, and then extraordinary things follow." Mr. Qian Xuesen promoted a comprehensive integrated seminar hall system (a mixture of expert system, machine system, and knowledge system) that is "people-oriented, man-machine integrated, from qualitative to quantitative (and back to qualitative)." The system is likely to be the golden key to solve the open and complex giant system problems.

In the development process of artificial intelligence, many researchers realize artificial intelligence in two ways: functional route and structural route. As everyone knows, the "structure" and "function" of intelligence include both fact part and value part. Generally, what is simulated is relatively objective factual "structure" and "function" and there is still nothing to do with the subjective value of "structure" and "function" The brain-like simulation should be a factual functional simulation, but it is still far away from the valuable "structure" and "function" As we know, although factual Boolean algebra has been widely used, the exploration of value Boolean algebra is far from beginning.

Many scholars discuss objective objects at two levels: "structure" and "function" They believe that "structure is the basis of function, and function cannot be realized without structure, but function is also the expression of structure, and each function is a specific structure that works" Although this method of classification has the advantage of facilitating analysis, it also splits the organic connection between "structure" and "function" (for example, brain-like is an analogous simulation metaphor that has both structure and function; splitting the two parts is just like splitting the yin and yang fish when talking about the "*Book of Changes*" or talking about "community" and "mutual" separately). Especially, good suanji are more reliable than good computers for the relationship between heterogeneous and supernatural powers. Being able to deal with multiple accidents in a timely and effective manner is the key. Furthermore, on the surface, the lack of generalization, abstraction, and dialectical ability is the pain point of current intelligence. In fact, how to effectively deal with various contradictions may be the main problem of intelligence.

The biggest feature of the structure and function of intelligence is the flexible combination of the "one" and "many" Individuality (individual intelligence) is the "one" while commonality (group intelligence) is the "many" and sometimes vice versa. In the context of incomplete data, lack of information, and insufficient knowledge, people can still compensate for sparseness, predict and analyze unknown trends, feel the incomplete cloze (filling in the blanks), perceive related clutter and irrelevance, and have a more in-depth situation awareness in the human-machine environmental system.

Intelligence is not composed of a series of isolated disciplines. The traditional intelligence classification does not actually conform to the profound nature of this discipline (just like the case in the classification of mathematics disciplines, where arithmetic is the science of studying numbers, geometry is the study of spatial objects, and algebra is the study of equations, and analysis is like studying functions). What really matters is not the nature of the objects studied but their mutual relationships among themselves and their relationships.

Swarm intelligence, the most common form of intelligence that is different from individual intelligence, covers all aspects of intelligence classification such as symbols, connections, and behaviorism. However, the proportion of factual symbols/connections/behavior contained therein will be relatively reduced, and value symbols/connections/the proportion of behavior will increase relatively. All intelligence can be divided into factual and value intelligence forms. Artificial intelligence is only a part of factual intelligence, while value intelligence is the very wisdom.

Mr. Qian Xuesen's systematic engineering thoughts are an attempt to effectively integrate Eastern thoughts and Western technology. From Monkey King's eye-catching eyes (contradiction between the amount of information and resolution) to Sherlock Holmes's magical computers (contradiction between facts and values), from "good-looking and useable" to "use it well" he organically integrates Tao and reason, name and philosophy, man and machine, state and trend, sense and percept, environment and environment. He has a long-term vision and insight into the smallest; what he studies is commendable, and it is the future of social civilization, in which, intelligence, this complex field, will play a leading role. However, it should be noted that the intelligence here is not artificial intelligence and machine learning, but the integration of man-machine (environment) intelligence, so Mr. Qian's systematic engineering thoughts can also be referred to as human-machine environment system (interactive) engineering thoughts system.

At present, in many AI-assisted decision-making systems, there are many obstacles. Because humans and machines are in different inertia/

coordinate systems, it is difficult for machines to keep up with the jumps, crossings, and speeds of human thinking. Then, how does the situation awareness in different inertial systems change? For example, the facts and values in the vision of a marshal, a soldier, and a machine are definitely different. The limited rational logic and awkward cross-domain capabilities of a calculation are the shortcomings of human-machine hybrid intelligence. The machine cannot understand the relationship of equality, especially the value of different facts. People can use irregular and incorrect methods and means (or use the famous banner of righteousness) to achieve formal and correct goals. People can also use ordinary methods to deal with complex problems, and they can also use complex methods to answer simple question.

The intelligent system generated by the future interaction will not only change various parameters, but it will also change various rules. Anyhow, the intelligence that only reflects facts can only be AI, and what can reflect both facts and values is the real intelligence.

2. The bottleneck problem of a human-machine hybrid

The famous British philosopher of science Philip Kitcher proposed a concept of "well-ordered" science to regulate what is a good science. A well-ordered science should include consultations among representatives of various viewpoints, scientists, decision makers, and ordinary people. Their dialogues should implement all the processes of science. The process of resource allocation, research methods, and the transformation of theoretical results into applications of scientific and technologic research involves the interests of everyone, and therefore it should also require everyone's voice.

Human-machine hybrid intelligence is basically a process of effective linkage between human intelligence and machine intelligence (AI) in response to changes in the external environment. The fundamental problem is also a problem of "good order" but this "good order" includes facts. Interaction sequence also includes value interaction sequence, including factual causal sequence (such as Pearl's causality) and value causal sequence (such as religious causality). Human and machine work together to complete a task A, which can be seen as a sequence composed of several subtasks (a, b, c, d ...). The requirements of these subtasks vary according to the changes in the external environment, including not only changes of existing constituent elements and changes in attributes, but also changes in themselves and their relationships. Simply speaking, there are both objective factual changes and subjective value changes. How do we organize these subjective and objective subtask sequences efficiently? In other words, how do we form a good order faster, better, and more skillfully?

Clear division of labor sequence is very important for human-machine integration. For example, a human deals with the direction sequence, while a machine deals with processing sequence, and more specifically, these direction sequences can also be divided into more specific problems: which direction sequences can be grasped by humans, which direction sequences are not easy to handle for humans, which process sequences can be processed by machines, and which process sequences are not easy for machines to process.

Knowledge is divided into perceptual knowledge (including perception, feeling, appearance) and rational knowledge (including concept, judgment, reasoning). Thinking refers to the rational knowledge based on perceptual knowledge, which is the generalization and sublimation of perceptual knowledge. Appearance is the image of a certain type of things reappeared in the mind and the bridge from perceptual knowledge to rational knowledge. Concept is the cell and main form of thinking. Benevolence is people's will. Righteousness is people's path. Tao is the law of nature. And morality is what people get naturally.

Human-machine hybrid collaboration oriented to DSA is the process of organically combining the perceptual and the rational, representation and concept, benevolence and morality in the group + individual to form a good order. It is a systematic engineering of (multiple) humans, (multiple) machines, and (multiple) environments, in which the algorithm of calculation is a programmable part, and the algorithm of suanji is a describable part. In addition, there are indescribable follow-up parts of subjective and objective hybrid, such as intuition and insight, half-truth and half-lie, half-trust and half-doubt, half-push and half-accept. How to achieve the well-ordered integration of these complex systems may be beyond the scope of existing mathematics and science.

Some people think that in modern China, science and technology all appeared as a kind of "advanced" and "civilized" role, bringing the Chinese people the idea that "science is good" and it has been influencing people until today. Subconsciously we may think that science = correction. We always see the "science" with some positive value. When we say, "this is not scientific" we mean "this is not right" In fact, science also has negative effects. Science and technology such as Freon and DDT have brought negative effects. So, what exactly is science and technology? Similarly, what exactly are AI, the internet, and the atomic bomb? Maybe they are a double-edged sword of Damocles hanging over humans.

It is a usual case that AI has "order" without "good" or "good" appears after "order" (it is hard to see "good" before "order"). The purpose of human-machine integration is to ensure the structure and function of "good" before "order."

Just as that a knife can cut vegetables as well as kill people, the problem of "good" and "order" that coexists with facts and values is still an ethical dilemma of AI. The integration of humans and machines is to make AI both "good" and "ordered."

3. Why is it hard for AI to land

Why is it difficult to for AI to land? Why is artificial intelligence often criticized? Some people say that this is caused by science fiction movies, science fiction novels, video games, news media, etc. This point of view has certain reasonable elements, but there is a more important fact that everyone has overlooked: the "human-machine environment" system integration intelligence is often mistaken for "artificial intelligence (or even some algorithms)."

Those things, no matter if they are military or civilian, no matter if they are automation products or intelligent systems, generally doing a better job in terms of safety, efficiency, and comfort, are usually grounded and accepted by everyone. To get these advantages, most human-machine environmental systems should be harmonious and consistent, at least not simple AI + certain field or certain field + intelligent algorithm.

The essence of intelligence is not data, algorithms, computing power, and knowledge. It is the source of living intelligence to emphasize the mechanism of generating data, algorithms, computing power, and knowledge. By analogy, the "Talents" cultivated by the existing artificial intelligence education system may still be "robots" without "souls" which is caused by the simple "algorithm" Having computing but not skills, having methods but no Tao, having feelings but no ignorance, reason but no emotion, state but no situations, (like) brains but no minds, appearance but no meaning, eyes but no beads, we can only move around in the circle of possibilities, but we cannot try to explore the impossible world. Even if there is some exploration, we can only jump on the riverbed of the family similarity, which is still far from being able to do anything about real nonfamily similarities.

In addition to the human-machine environment system interaction, the second aspect is the understanding and acceptance of in-depth situation awareness. For example, in many situations, we only know the registration and correction between time and space, but do not know those of the state, situation, sense, and percept; we only know the distortion solution of the noncooperative distance, but we forget the fuzzy expansion of the cooperative distance; we only know the frequency conversion and variables, but do not think about abnormalities, changes, knowledge, and flexibility; we only know the data link, information chain, but we do not consider the fact chain and value chain, even the human-machine environment system chain formed by the entanglement of the state chain, the potential chain, the sense chain, and the knowledge chain; we only know the homogeny, uniform, and sequential situation awareness single modulation, ignoring the more important heterogeneous, nonuniform, random situation awareness multilevel arrays, as well as the rapid mobility of the first and the second knowledge and the accurate flexibility of the second and the second sense, as well as the self-reliance between the state, the situation, the sense, and the percept; we only know the human model with machine's functions, but we do not know the machine model with human's ability; we only know the simulation verification structure, but we do not pay attention to the function obtained in actual combat.

When something happens, we will consciously or unconsciously associate it with things that have just happened or leave a deep impression on us from time to time, and we establish our own personalized "causal relationship" situation spectrum (not only the map). Those things that are indeed relevant are called objective factual connection, those speciously relevant called a possibility connection, and the irrelevant ones are called a subjective intentional connection. These connections that often occur in life are all part of intelligent cognition. The objective factual correlation part of the transformation that can be programmed is often called AI, but the possibility correlation and the subjective intentional correlation are filtered out, which are the important components of the flexibility of personalized intelligence.

In short, the reason why it is always difficult for AI to land may be the following three problems. First, AI algorithms are used to simplify the problem of a complex system in the human-machine environment. Second, people only pay attention to situation awareness without focusing on the DSA. Third, people always ignore the wormhole connection between irrelevant things.

4. Side effects of intelligence (including AI)

Every coin has two sides, so do mobile phones, and so is intelligence. When the human-machine environment system dynamically interacts (generates intelligence), the ways, methods, schemes, means, and the tools will be adjusted and recombined appropriately due to changes in time, space, objects, attributes, relationships, conditions, rules, emotions, states, trends, perceptions, etc. It can be said that law changes as times change. What intelligence needs to solve is often the problems people face in real life, such as security threats, efficient processing, and accurate prediction. Intelligence contains past experience and data, but it does not rely solely on these pasts. It also contains future influences on the very moment, such as expected feedback. Generally speaking, intelligence that cannot adapt to changes is not true intelligence.

For physiologic diseases, we need to prescribe the right medicine, and it is the same for intelligence: there is no universal intelligence that can cure all problems. All forms of intelligence and cognition have scopes. As those that can cure all diseases are fake medicines, the omnipotent intelligence is fake intelligence. Even true intelligence has side effects, just as an old saying goes, "Cleverness is mistaken by cleverness." So true intelligence has its shortcomings, but these shortcomings are different from the shortcomings of automation. The biggest feature of the shortcomings of intelligence is that they can be repaired and improved autonomously and timely, while those of automation cannot be so, which is a bit like spilled water that cannot be gathered up.

The constant human-machine environment interaction and changes determine that there is no recognition task mode in the world. "Oranges born in the south are oranges, but oranges born in the north are Trifoliate orange." Such examples are also not uncommon in the field of intelligence. Machine's intelligence can assist people's learning, reasoning, and decision-making, but it can also interfere with people's reasoning and judgment. The case that "good intentions" do bad things not only exists among humans but also between humans and machines, such as the active takeover of AI assistants problems and so on. Furthermore, due to the inexhaustible complexity of problems, things that are unexplained, difficult to explain, unexplained, and inconvenient to explain are everywhere. Even humans are unpredictable, let alone the fuss. Algorithms designed in different situations are used to influence, verify, correct, and change people's intuitions and thinking results in real time. If it is trivial, it does not matter much; but for those big events that determine the destiny of mankind, who dares take this responsibility if there are errors and mistakes?

AI can help people, hinder people, and destroy people in human-machine hybrid interaction, human-machine mixing, human-machine hybrid intelligence, etc. When doing these jobs or applying for projects, I hope the reviewers and managers can also objectively see its bad side besides the good side of AI. In many situational tasks, the probability of being bad may be higher.

If there is a clear standard answer, it should not be called intelligence. Intelligence (including AI) is a moving lamp, with light and shadow. Be careful!

5. A piece of advice for human pilots

In late August 2020, an aircraft with human, the "dog fight" test, ended in the DARPA's (Defense Advanced Research Projects Agency) drone battle. After the excitement, the key to AI's victory lied in its strong offensiveness and shooting accuracy from the review after the test. Its main problem is that there are errors in judgment. According to US military tests, the tested AI system often makes mistakes in basic fighter maneuvers. AI has turned the aircraft more than once in the direction it thinks the human opponent's aircraft will go, but it has been proven to have misjudged the idea of the human pilot many times. It is not difficult to understand this: even human pilots often make mistakes in judging the intentions of opponents, and AI systems may do worse due to the lack of the ability to understand creative tactics, so it is not surprising to make such mistakes. However, an AI is still able to maintain the advantage over human pilots as a whole, and its calculation system ultimately prevails in the entire confrontation due to its "excellent aiming ability" and the ability to track opponents' aircraft.

In short, unmanned aerial vehicle AI takes the lead in the accuracy of "state" and the speed of "sense" but it does not yet have advantages in the judgment of "situation" and the prediction of "percept" It is suggested that the pilots in the aircrafts with humans should work hard on fake actions (like Jordan, Kobe, and James) and breaking rules (like Sun Tzu and Zhuge Liang). Without rules, all algorithms and (mathematical) models will lose boundaries, conditions, and constraints, and all computers will no longer be accurate and reliable. When the probability formula changes from calculation to fortune-telling, the advantages of machines may not be as good as humans.

Humans make value-based decisions—discussing major issues rather than just calculating gains and losses—but machines make factual decision—discussing gains and losses, not justifications. The relationship between state and trend and between sense and percept are respectively quantitative and qualitative. The "situation" is the greatest possibility within a certain period of time. Whatever is in the "situation" is not already in the "state" what is in the "percept" is not already in the "sense" It can be said that a single spark can start a prairie fire. If the goal is clear, the opponent should be or can only be the corresponding system in a game with a large system composed of control units and equipment, but not the person who operates the equipment or the person who designs and controls the system. In this regard, people have great weaknesses. The key point is that the dynamic changes of the long-, medium-, and short-term goals in the development environment will cause the goals to be ambiguous and even vague.

The current artificial intelligence, like a high-speed rail, is fast, but it needs a track. But the real intelligence should be like an airplane, it does not need a specific track and route as long as it can reach the destination. The errors of situation awareness can be divided into errors of state, trend, sense, and percept, and they can also be divided into factual/value errors. The application of artificial intelligence in weapons is mainly reflected in the machine-to-machine task arrangement and the real-time re-targeting of weapons. This sort of effect prioritization of typical "service providers" will be executed at the tactical level, depending on whether the intelligent machines can digest and analyze data from the entire battlefield. In fact, the factual and valuable data, information, knowledge, responsibilities, intentions, and emotions in the distribution of human-machine functions will be the focus and difficulties of future man—unmanned confrontation.

The most terrifying thing about humans is not lacking thoughts but full brains with standard answers; the most terrifying thing about machines is having "thoughts" and being able to "lose before it wins" the most terrifying thing about a human-machine hybrid is being mistaken for artificial intelligence.

CHAPTER 4

Three hurdles through which artificial intelligence cannot go

Contents

With the rapid development of artificial intelligence, many disciplines are slowly being cross-integrated. After three ups and downs of artificial intelligence, its flaws and limitations are being revealed.

1. Interpretability: the first hurdle through which artificial intelligence cannot go

Nowadays, the interpretability of artificial intelligence is becoming a hurdle. In 2019, the European Union issued the "Artificial Intelligence Code of Ethics" which clearly stated that the development direction of artificial intelligence should be "trustworthy" including security, privacy, transparency, interpretability, and other aspects.

Artificial intelligence applications aim to output decision-making judgments. Interpretability refers to the degree to which humans can understand the reasons for a decision. The higher the interpretability of the artificial intelligence model is, the easier it is for people to understand why certain decisions or predictions are made. Model interpretability refers to the understanding the internal mechanism of the model and understanding the results of the model, whose importance is reflected in the following aspects. In the modeling stage, it assists developers in understanding the model, making comparisons and selections of models, and optimizing and adjusting the model when necessary. In the stage of operation, it helps to explain the internal mechanism of the model to the decision-making party

Integrated Human-Machine Intelligence
ISBN 978-0-323-99562-7
https://doi.org/10.1016/B978-0-323-99562-7.00004-8

and interpreting the results of the model. For example, the decision-making recommendation model needs to explain why a certain solution is recommended for this user.

At present, the understanding and definition of artificial intelligence in various fields differ because of the field distinction, but there is consensus in common technologies and basic research. The first stage of artificial intelligence aims to solve problems and carry out logical reasoning through machine theorem proofs, expert systems, etc. In the second stage, it realizes environmental interaction, which obtains information from the operating environment and exerts influence on the environment. In the third stage, it moves toward cognition and thinking ability, discovering new knowledge through data mining systems and various algorithms.

Strictly speaking, the United States' artificial intelligence technology is generally in the lead of the world, but when it comes to human-machine hybrid intelligence, it seems not so advanced, or it even lost its leading position (maybe there is no generation gap at all between China and the United States in terms of intelligence). The reason is the human. For example, in this epidemic, the United States should be much better in terms of medical software and hardware and medical personnel. Unfortunately, as it goes in the "Trisomy" "Weakness and ignorance are not obstacles to survival, but arrogance is." The errors and mistakes of the leaders have greatly compromised many advancements, or even lost them. This reminds people that it may be a similar case considering a recent report in the US Arms Control Journal. The US Department of Defense requested US$28.9 billion for the modernization of US nuclear weapons facilities in the 2021 fiscal year to support the Trump government's strategic development focus: increasing the high degree of automation of nuclear command, control, and communications (NC3) infrastructure, increasing its speed and accuracy. But it also raises a disturbing question: what role will artificial intelligence autonomous systems play in determining the destiny of mankind in future nuclear wars? At present, calculation-aided decision-making is still in its infancy, which is prone to unpredictable failures. Although machine learning algorithms are good at specific tasks such as facial recognition, they also have inherent "bias" conveyed through training data. Therefore, it is necessary to adopt a cautious and responsible attitude in applying artificial intelligence to nuclear weapons accusations. As long as nuclear weapons exist, humans (not machines) must exercise ultimate control over the use of nuclear weapons. At this time, the true ability of man-machine integration of intelligence will be as important as that in the epidemic control.

The integration of human and machine intelligence is fundamentally the representative of the combination of science and technology, humanities and art, mathematical symbolic fact language and natural experience value language. Time and space can not only bend in the physical realm, but they also can be distorted in intelligence. We can say that philosophical logic has experienced a turning on the world's origin problems and research methods, and the 20th century analytical philosophy, the analysis of human language tools, has become a philosophical "revolution" in human thought symbolized by Wittgenstein, which directly induced the rapid development of artificial intelligence technology represented by the Turing machine and the Turing test. However, Jin Guantao's "Philosophy of Authenticity" believes that in the 21st century, analytic philosophy, which has in fact caused the imprisonment of thought, has finally confined philosophy in a cage; symbols can have their own authenticity when they do not refer to objects of experience, and this conclusion can be established for both mathematical language and natural language. At the same time, the authenticity of pure symbols can be embedded in the authenticity of experience. Furthermore, scientific research and humanistic research can become two areas that are unified but do not overlap with each other and have their own authenticity standards. The great progress of mankind is to make the authenticity instinct (the objectivity of common sense) under the control of the system of ultimate care and corresponding value. But today, the two pillars of authenticity are being overturned by scientific progress, and a truly terrifying thing has appeared: human beings are irresistibly reduced to intelligent "animals." In a world where there is no distinction between true and false, there is neither righteousness and wrongness, nor a real sense of morality and dignity of life.

Humans do not describe the world with only symbolic equivalence or contain logical relations. Education for humans is not equal to learning and knowledge but to inducing desires in a good direction. The calculation itself cannot cross the "understanding" gap, while the human is the only one who can cross the embarrassment of symbol pointing. For the subject, signs and experience are mixed, logic and nonlogic are mixed, axioms and nonaxes are mixed; what is more, data, information, and knowledge are mixed together. This is the main reason why interpretability is difficult. The man-machine hybrid is how symbols (mathematics) are embedded in the subject's experience (controlled experiments) to varying degrees, just as Lao Tzu said in "*Tao De Ching*", "Tao is the Originator of undivided universe, the universe gives rise to Yin and Yang; Yin and Yang give rise to heaven, earth, and people, and they give rise to all the worldly things."

The so-called artificial intelligence is to a large extent just the use of the calculation's ever-increasing computing power, and it is doomed to be wrong to be taken this way. Humans learn and use according to the situation, while machines are rigid under all circumstances. Human intelligence is the ability to perceive a small sample of situation awareness. A well-known example of situation awareness is the looking, smelling, asking questions, and exploring the pulse in Chinese medicine. The difference between natural language and mathematical language is used to break the divergence between mind and physics, thereby unifying facts and values.

Situation awareness originates from the "61st difficulty" in the "Difficulty Sutra" which says, "In the Scriptures, one who can know a thing after seeing it is a god; one who can know it after hearing it is a sacredness; one who can know it after asking questions on it would be a worker, and he would be a smart one when he just explores the pulse and know the disease." What is it? The earliest use of the four-character joint name of "looking, smelling, asking questions, and exploring the pulse" should be in the "Ancient and Modern Medical System": "the four characters of looking, smelling, asking, and exploring the pulse (four characters in Chinese) is truly the program of Chinese medicine." "Looking" is to observe the patient's development, complexion, tongue coating, expression, etc. "Smelling" is to listen to the patient's voice, coughing, wheezing, and "smelling" is to smell the patient's bad breath, body odor, and other odors. "Asking" is to ask about the symptoms felt by the patient, previous illnesses, etc. And "exploring the pulse" is to probe the pulse or examine whether there is a lump.

The fundamental reason the interpretability of artificial intelligence is problematic is that it contains not only mathematical language but also natural language, and even thinking language (so it is impossible to cross this hurdle). Human-machine hybrid intelligence cannot only suspend the subject, but it also changes the subject with ease. We can probably go straight to the purpose and intention realization when real-time and timely in-depth situation awareness can be achieved in the interaction of human-machine- environment systems, and signifier, the signified, and the signification could be organically switched with mathematical language, natural language, and thinking language.

2. Learning: the second hurdle for artificial intelligence

Human learning is not knowledge but the method of obtaining data, information, knowledge, and experience. And machine learning is getting data, information, and knowledge.

There are similarities among different material systems, while each subsystem of the same material system and the overall system also have similarities, and material systems with different forms of motion and properties obey similar physical laws. These facts all show that similarity is an essential characteristic of nature. For example, the mechanical system composed of mass-spring-damping and the circuit system composed of resistance-inductance-capacitance are similar systems, reflecting the similar relationship between physical phenomena (in general, similar relationships can be used to simplify complex systems for research). It is easier for machines to learn and transfer this homogeneous and linear similar system. However, it is challenging to realize the analogy and conversion of heterogeneous and nonlinear similar systems. However, human learning can freely gallop and wander between symmetry and asymmetry, homogeneity and nonhomogeneity, linearity and nonlinearity, homology and nonhomology, isomorphism and nonisomorphism, empathy and nonsympathy, sympathy and nonsympathy, periodic and not periodical, topology and nontopology, family and nonfamily.

Machine learning is inseparable from time, space, and symbols, while human learning is a system that changes with changes of value, facts, and emotions; machine learning follows and relies on existing rules, while human learning studies how to modify old rules, break conventional rules, and establish new rules. For example, the genuinely outstanding leaders and commanders are trying to break the rules—reform, rather than decline and decay step by step steadily, let alone watching the epidemic spread, but staring at the campaign and the black hat.

On March 16, 2017, the Defense Advanced Research Projects Agency (DARPA) planed to launch the "Lifelong Learning Machines" (L2M) project, which aims to develop the next generation of machine learning technology and use it as a basis to promote the third time AI technology wave. DARPA believes that the development of AI technology has gone through the first and second waves and is about to usher in the third wave. The first wave of AI technology was characterized by "rule knowledge" Typical examples are the Windows operating system, smartphone applications, and programs used by traffic lights. The second wave of AI technology is characterized by "statistical learning" and typical examples are artificial neural network systems, and progress has been made in areas such as driverless cars. Although the AI technology mentioned above has strong reasoning and judgment ability for apparent problems, it cannot learn, and the ability to deal with uncertain problems is also weak. The third wave of

AI technology will be characterized by "adaptation to the environment" AI can understand the environment and discover logical rules, to train itself and establish its decision-making process. It can be seen that AI's continuous self-learning ability will be the core driving force of the third wave of AI technology, and the goal of the L2M project is precisely in line with the characteristics of the "adapt to the environment" of AI in the third wave. By developing a new generation of machine learning technology to learn from the environment and sum up general knowledge continuously, the L2M project will lay a solid technical foundation for the third wave of AI technology. At present, L2M includes a massive base of 30 performance groups, and works through grants and contracts of different durations and scales.

In March 2019, researchers from the University of Southern California (USC), a partner of DARPA, published their research on exploring bionic artificial intelligence algorithms. Francisco J. Valero, L2M researcher and professor of biomedical engineering and kinematics at the USC Viterbi School of Engineering Cuevas, joined the college's doctoral students Ali Marjaninejad, Dario Urbina-Melendez, and Brian Cohn to publish a paper in "*Nature Machine Intelligence*" which detailed the successful development of robotic limbs controlled by artificial intelligence. The limb was driven by animal-like tendons that can automatically recover from balance disorders.

What drove USC researchers to develop this robot limb is a bionic algorithm that can learn walking tasks autonomously within just 5 minutes of "unstructured play" In other words, they made a robot learn its own structure and surrounding environment by performing random movements.

Current machine learning methods rely on preprogramming of the system to handle all possible scenarios, which are complex, workload-intensive, and inefficient. In contrast, USC researchers revealed that artificial intelligence systems could learn from relevant experience, and they strive to find and adopt solutions to meet challenges.

In fact, for much infinite learning, it is difficult for people to achieve lifelong learning. There is always some knowledge people can learn, and some other people know a little or even nothing. For the machines lacking "common sense" and "analogy mechanism" lifelong learning maybe a slogan. The first thing that needs to be clarified should be what can be learned and what cannot be learned?

Human learning is omnidirectional learning from different angles. One thing can become multiple things, one relationship can become multiple relationships, and one fact can become multiple facts and multiple values. What is more interesting is that sometimes people's learning can turn many different things into one thing. Many different relationships can become one; multiple facts can become one factor even one value. Moreover, machine learning is essentially the cognition of people (one or some people). In the strict sense, this is a kind of "self-righteousness" for people can only recognize things they are used to or familiar with. Therefore, the limitations and narrow-mindedness of this group of people are also unconsciously integrated into the model and program. Therefore, this one-to-many transformation mechanism is often inherently inadequate from the beginning. Of course, machine learning is not useless. Although it is not suitable for intelligence, it should be good for calculation or automation applications.

If the essence of learning is classification, human learning is a method of obtaining and creating classification, while machine learning is simply a classification method. DARPA's "Lifelong Learning Machine" project may be essentially a beautiful bubble that floats up and down in the air when it blows. Although it will be colorful under the sun, it will eventually burst.

3. Common sense: the third hurdle for artificial intelligence

Like all medicines, all knowledge has scopes and premises. Without these, side effects of knowledge will emerge. Knowledge is only the material and raw material of common sense. Machines have only "sense" but no "comprehension" and they cannot unite knowledge and action. Knowledge should not be attached to thought but should be integrated with it. It is best to abandon it if it cannot change thought and make it perfect. Having knowledge without ability of knowing how to use it is worse than learning nothing; such knowledge is a dangerous sword that will cause trouble and harm to its owner. One of the most effective ways to limit the side effects of knowledge is the formation of common sense. Generally speaking, common sense is often fragmented, and situation awareness is formed through the perception of these fragmented common sense states and trends, some kind of very conscious knowledge and insight. In addition, common sense is an essential ability for humans to perceive and understand the world. A typical AI system lacks a general understanding of the operation of the physical world (such as intuitive physics), a basic understanding of human motivation and behavior (such as intuitive psychology), and an adult's cognition of universal things.

While developing the second-generation artificial intelligence technology and its military applications, DARPA is actively deploying the development of third-generation artificial intelligence. In the 2018–20 fiscal year, it was committed to the basic research of third-generation artificial intelligence through new projects and continuation projects. It aims to break through artificial intelligence's basic theories and core technologies through research in machine learning and reasoning, natural language understanding, modeling and simulation, and human-machine hybrid. Related projects include "machine common sense" "lifelong learning machine" "explainable artificial intelligence" "reliable autonomy" "active interpretation from different sources" "automatic knowledge extraction" "ensure the reliability of AI antideception" "accelerated artificial intelligence" and "foundation" "artificial intelligence science" "machine general perception" "learning with less data" " knowledge-oriented artificial intelligence reasoning model" "advanced modeling and simulation tools" "complex hybrid system" "human-machine communication" "human-machine symbiosis" etc. In addition, DARPA recently released a wide range of institutional announcements on artificial intelligence basic research projects. It also includes "artificial intelligence and learning science for opening the world's singularity," "human-machine hybrid collaboration social intelligence team" and "real-time machine learning."

If learning cannot teach us how to think and act, that would be a great pity. Because learning is not used to make those people without thoughts have thoughts and make the blind see. The job of learning is not to provide a vision for the blind but to train and correct vision, as far as they have the vision and can be trained. Learning is good medicine, but any good medicine may deteriorate, and the expiry date depends on the quality of the medicine bottle.

The main achievement of Vladimir Voevodsky was the development of a new algebraic cohomology theory, which provides a new perspective for the profound theory of number theory and algebraic geometry. The characteristic of his work is that he can handle highly abstract concepts quickly and flexibly and use these methods to solve fairly specific mathematical problems. The concept of cohomology was initially derived from topology. Topology can be roughly said to be the "science of shape" where examples of Wowo shapes such as sphere, torus, and their high-dimensional analogs are studied. Topology studies the fundamental properties of these objects that remain unchanged under continuous deformation (no tearing allowed). In layman's terms, cohomology theory provides a way to divide topological objects into pieces that are easier to study. The cohomology group contains information on how to assemble these essential pieces into the original object. The main object of research in algebraic geometry is algebraic clusters and the public solution set of polynomial equations. Algebraic clusters can be represented by geometric objects such as curves or surfaces, but they are more "rigid" than deformable topological objects.

The "Mosaic War" concept proposed by the DARPA Strategic Technology Office in 2017 believes that the future battlefield is a mosaic of low-cost, low-complex systems connected in many ways to create an ideal fit for any scene interweaving effect. A part of this concept is to "combine existing weapons in new and surprising ways" focusing on manned/unmanned marshaling and decomposition capabilities, and allowing commanders to seamlessly summon sea, land, and air capabilities based on battlefield conditions, regardless of which unit is providing combat capability.

Simply put, the "mosaic warfare" and "machine common sense" introduced above are both new topological systems against the game human-machine environment system, just like the "Motivic Cohomology" theory created by Wowski. Actually, the most powerful are not those basic knowledge, regulations, and rules, but those who apply this basic knowledge and these regulations and rules to obtain universal success in practice.

4. The first principle of intelligence

Hume believes that "All sciences are related to human nature, and the study of human nature should be the foundation of all sciences." Any science has something to do with human nature. Every approach returns to human nature again no matter how far it seems from the human nature. This is the case for science, and the complex system that contains science is no exception. The actual intelligence has a dual meaning. One is the formal meaning of facts, that is, the logic of rational action and decision-making in general, and how to choose rationally when resources are scarce and maximize the utility. The other is the substantive meaning of value. It neither presupposes rational decision-making nor presupposes scarcity conditions. It only refers to how humans plan from their social and natural environment. This process is not necessarily related to utility maximization but to a perceptual category. The power of reason is limited because human behavior is not only influenced by acceptability, but it also has an "irrational" side in the real world. The "ethical design" of artificial intelligence is probably a daydream. The reason is simple. Ethics is still a complicated system that is difficult for people to observe. Simple ethical rules are often the most difficult to achieve. For example, people who are in difficulty should be helped. This ethical rule is brutal (and those who follow it can be easily deceived). For AI is a tool, so ethical design should have more sci-fi elements than science and more imaginative elements than natural elements.

The current artificial intelligence and future intelligent scientific research have two fatal shortcomings: ① equating mathematics with logic and ② confusing the sign with the object's reference. Therefore, the difficulty and bottleneck of the in-depth situation awareness of human-machine hybrid lies in the following aspects: ① the nonsymbolic (variability) of (symbolic) representation, ② the illogical (nonauthentic) nature of (logical) reasoning, and ③ the nonlogical (objective) decision-making objectivity (subjectivity).

Intelligence is a complex system that includes both calculation and sunaji. Generally speaking, artificial (machine) intelligence is good at calculating objective facts (truth), and human intelligence is good at the suanji of subjective value (rational reason). When the calculation is greater than suanji, artificial intelligence can be emphasized; when the suanji is more significant than calculation, human intelligence should be selected; when the calculation is equal to suanji, it is better to use human-machine intelligence. Feynman said, "Physicists are only trying to explain events that do not depend on chance, but in the real world, most of the things we try to understand depend on chance." However, the core of the intelligence of both humans and machines lies in change: change with time, change with circumstances, change with law, change with the situation.

How do we realize the suanji system after the hybrid of suanji (experience) of human and calculation (model) of machine? Taiji Bagua Diagrams (Eight Diagrams) is a typical system of jisuanji (calculation + suanji). The logic of "and-or-not" is based on both human experience and material (machine) data, that is, human valued "and-or-not" + machine factual "and-or-not" One of the tasks of human-machine hybrid intelligence and deep situation awareness is to open up the narrowness of AND, OR, and NOT. The empirical probability of humans is different from the substantial probability of machines. It is a value probability that can penetrate the barriers of nonfamily similarity and use the success or failure results of other fields to affect the situation awareness in the current field, such as sympathy and empathy, empathy and trust, and so on.

The core of human intelligence is the object pointed to by intention, the core of machine intelligence is the object pointed by signs, and the core of human-machine intelligence is the combination of the object pointed by intention and the object pointed by signs. They are all related to existence, and existence is divided into factual existence and value existence, as well as responsible existence. For example, when the exact epidemic exists, Academician Zhong Nanshan talks about objective existence, and Trump talks about value. At the same time, what they talk about includes responsibility, but one is scientific responsibility, and the other is a political responsibility.

Generally speaking, mathematics solves the problem of equivalence and compatibility (inclusive). However, the equivalence and compatibility (inclusive) of this world are very complicated. The objective factual equivalence and subjective value equivalence are often not the same. On the contrary, objective factual compatibility (including) and subjective value compatibility (including) are often not the same things. Therefore, the world should be composed of facts and values. In other words, the world is composed of the nonnumeric part in addition to the mathematical part. Science and technology are relatively rational, based on mathematical logic (axiomatic logic) and experimental verification. The humanities, art, philosophy, and religion are based on nonnumeric logic and the relative sensibility of imagination. Partly, the combination of the two enables human beings to exist endlessly in nature.

In a sense, mathematics solves the intelligent "being" (is, existence) knowledge (such as equivalence, inclusion problems). However, it is far from or even impossible to solve the "should" (response, righteousness) problem. For example, when natural philosophers tried to find the eternal and unchanging origin in the ever-changing nature, one of them, Parmenides, discovered that no natural thing is eternal and unchanging; the only thing that is genuinely unchanging is "exist" In a judgment (S is P), both the subject and the object can change, and the only thing that does not change is the "being" (being). In other words, everything is "is," and all "exist" but the things in it will one day be "not" or "nonexistent" but "yes" or "existence" will not happen because of the changes in the birth and death of things. Change is eternal and unchanging; this "is" or "being" is the basis for making things "being" or "being" Therefore, unlike the cosmology that explores the original cosmology before time, Parmenides mainly asks about the prior existence in logic. Although it is not yet, it is equivalent to what people call "essence" and part of this "yes" may be mathematics.

The relationship between human and machine environment has both directional closed loop and nondirectional open loop, or both directed open loop and nondirectional closed loop. Most autonomous systems are directed closed-loop behavior. The computing system integrated with the human-machine hybrid environment system may be a secret channel to solve Hume's question; that is, through the combination of human computers and machine computers, a "qualitative leap" from "facts" to "value" has been achieved.

Some people think that "the wisdom of the whole scene is a great integration of technology." In fact, this refers to an aspect of engineering application. If we look into it in depth, it is still a great integration of the character and environment system of science and technology, humanities and art, philosophy, ethics, customs, and beliefs, just like this fight against the epidemic. A better human-machine hybrid interaction relationship is like a yin—yang map. You have me in you, and I have you in me. They are interdependent and balanced. Just like the current relationship between China and America, if the United States wants to remove Huawei's chips, Intel will suffer damage. In a nutshell, the current human-machine hybrid relationship is two fishes, connected head to tail, black and white.

Every thing, every person, every word, every letter can be regarded as a congregation of facts + values + responsibilities. Psychologic feedback is different from physiologic and physical feedback. The logic of sensation is different from the logic of percept. For sense, concepts are figures; for percept, concepts are signs. From the perspective of intelligence, there is no so-called "meta" only changing "meta" The "meta" can be a very big thing. For example, the solar system and the Milky Way can be regarded as a meta unit; people call it Smart Mijuko.

Scientists often only try to explain events that do not depend on accidental events, but in the real world, the things that human-machine environment systems often try to understand are mostly determined by accidental factors, just like the fate of human beings. Wittgenstein once commented on this, "Under the foundation of the entire modern worldview, there is an illusion that the so-called natural law is an explanation of natural phenomena." Kitcher has also been trying to resurrect a single event with reasons. However, an infinite number of things can affect an event. Which one should be regarded as the cause of it? Furthermore, science can never explain any moral principle. There seems to be an insurmountable gap between the question of "being" and "should" Perhaps we can explain why people think that certain things should be done or explain why humans have evolved to believe that certain things should be done and other things cannot be done, but going beyond these biologically based moral laws is still an open question for us. Professor Penrose of the University of Oxford also believes, "You cannot hear the same rhythm of 'tick' in the universe. Some things you think will happen in the future may have happened in the past. The order of occurrence of the two unrelated events in is not fixed in the eyes of two observers; that is, Adam might say that event P happened before event Q, and Eve might argue that event P happened after event Q. In this case, The clear sequence that we are familiar with (the past triggers the present, and the present triggers the future) completely disintegrated. That is right, in fact the so-called causality is completely disintegrated here." It may be the only thing lasting forever, which exists before this world and will also exist in the organizational structure of this world itself, is called "change."

In a sense, intelligence is the product of culture. Every concept and knowledge of human beings is dynamic, and only in practical activities can it produce multiple wormholes related to other concepts and knowledge and realize the state of "living" and the trend of "birth" At the same time, these concepts and knowledge will maintain a certain degree of stability and inheritance to maintain the immutability of genes in continuous evolution. Time and space are all possible conditions as knowledge concepts, and at the same time, they are also the limitation of many principles; that is, they cannot be completely consistent with the existing nature itself. The key to possibility lies in the premises and conditions. Generally, people often pay attention to the possibility but neglect to pay attention to its constraints and scope. We limit ourselves to those principles that are only related to categories, related to categories, and many principles that are not related to categories get no attention nor concern. In fact, the human-machine environmental system has flexibility in situation awareness. One problem with the purely physical concept of the mind is that it does not seem to leave much room for free will. If the mind is completely governed by the laws of physics, its free will is like a rock that "decides" to fall to the center of the earth. All intelligence is related to the human-machine environment system. The advantage of artificial intelligence lies in stitching, and the disadvantage of artificial intelligence lies in fragmentation. Pure artificial intelligence software and hardware that do not consider people and the environment are like those people who carve boats for swords, blind people touch elephants, and Cao Chong weights the elephant. In other words, it is automation.

Human learning is an interactive learning structure triggered by the initial indoctrination and the later environment (more important), and the machine lacks the ability of the later stage. Human learning is a mixed learning of facts and values, and it is a dynamic learning of weight adjustment. Human memory is also self-adaptive, changing with the human-machine environment system, and from time to time, it will find features that have not been noticed before. Through learning, people can turn the situation into momentum and influence into knowledge, and machines seem to be able to do it, but most of them are free from the "dead" situation of environmental changes and "stiff" knowledge. Sometimes it is a human factor, and sometimes it is a factor of environmental changes. People live in a complex system in which there are many variable subjects and variable objects interacting with each other. There are multiple links in human-machine hybrid. Some are suitable for humans, some are suitable for machines, some are suitable for humans and machines, and some are suitable for waiting for tasks to fluctuate. How do we determine these divisions and matching are very important? And how do we perceive in the situation ? Or how we generate a situation in a bunch of perceptions? How is the state, trend, sense, and percept in the time dimension? What are the state, situation, sense, and percept in the spatial dimension? What are the state, trend, sense, and percept in the value dimension? All the above aspects are very important.

So, how do we achieve a directed human-machine hybrid and deep situation awareness? One is the directionality of "universal facts" such as the directionality of the quantities rules, statistical probabilities, constraints, etc., in chess and Chinese chess, the algorithms used in human learning, machine learning, and the directionality of rational derivation, etc. Although the problems here are quite different, they all have only two directions (positive and negative), and the angle between them is not large, so it is called "universal-factual" directional. This kind of directionality, which is widely used in mathematics and physics, facilitates computing. The second is the directionality of "pan-value" that is, the directionality that we often use in our subjective intentionality analysis and judgment, but it is not convenient to measure it. We know that the vector here has an infinite number of directions, and the addition of two vectors with different directions usually results in a vector with different directions. Therefore, we call it the "pan-value" vector. This kind of "universal" directed mathematical model has too many directions for us, and it is inconvenient to apply it.

However, it is precisely because "universal value" has vector additivity and "universal thing" directional duality that it enlightens us to study a kind of cognitive quantity that has both duality, directional, and additivity. The directed distance of one-dimensional space, the directed area of two-dimensional space, the directed volume of three-dimensional space, and even the general N-dimensional space are all examples of such geometric quantities. Generally, people refer to directional metrics as directed metrics. Situation awareness is generally the directionality of "pan-facts" potential is the directionality of "universal values" sense is generally the directionality of "universal facts" and knowledge is the directionality of "pan-values" The human-machine relationship is a bit like quantum entanglement, and it is often not a question of "being or not" but of "being and not." and "not" exist in each other, the "being" can be calculated, the "no" can be deliberated, and the "being and not" can be calculated with suanji. Therefore, there must be a human staff and a machine staff in the future military human-machine hybrid command system, in which one is responsible for the calculation of "being" while the other is responsible for the suanji of "nothing" forming a system of "jisuanji" Not only can we grasp things intuitively, but we also understand the laws indirectly.

The science developed in the West focuses on the search for truth, and it is often divided into two categories: theoretical science and practical science. The purpose of the former kind is to handle knowledge and truth, while the latter seeks to control objects through human actions. The two categories are specifically manifested in the search for a system of proof of truth: truth in the formal sense (instrumental theory, logic), truth in the positive sense (physics, empirical world), and truth in the critical sense (post-physics, metaphysics). Mr. Yu Wujin believes, "The history of Western metaphysics" so far is composed of the following three reversals. First, the "subjective metaphysics" represented by Descartes, Kant, and Hegel reversed the Platonic "metaphysics of presence" Second, within the metaphysics of subjectivity, the metaphysics of will represented by Schopenhauer and Nietzsche reversed the metaphysics of rationality represented by Descartes, Kant and Hegel. The third is the later "The metaphysics of the fourfold whole of the world (Heaven, Earth, God and Man)" by Heidegger reverses its earlier "Dasein Metaphysics." Through these three reversals, we can draw the conclusion that intelligence is a kind of human-machine environment system interaction, which not only involves the research of rationality and logic, but also includes perceptual and illogical immersion. The current artificial intelligence is only an automated system that mixes human cognitive mechanisms with statistical probabilities, and it is far from entering the field of real intelligence. To achieve truly intelligent research, it is necessary to go beyond the existing artificial intelligence framework and honestly combine the Western "truth" with the Eastern "Tao" to form system of the calculation combining sith suanji, in which facts and values, human intelligence and wit, narration and proof, calculation and suanji mix with each other.

As Mr. Li Chaodong said (originally published in his paper "Northwestern Normal University" 2000, No. 5), "Western philosophy is a system of proof of truth, not a system of moral values, not the language of reason. It is a kind of opinionated truth language of 'legislation' for the world falling into the thought and experience, which is a system of fact judgment, not a system of value judgment. The truth that Chinese people pursue is related to 'right' and 'wrong' 'good' and 'bad' 'yes' and 'no'". In other words, Chinese people have never entered the field of truth or philosophy in the whole of proof system. Westerners speak of truth, and Chinese speak of reason. Reason is the law of nature, which describes what it is, neither self-righteous nor self-contained. The so-called "Tao follows nature" means that nature itself describes what it is. But the "being what it is" refers to the yin and yang, the heaven and the earth, the way of men and women, so we say the "yin and yang" is the "Tao" "Tao" is connected to the world, and the way of men and women is called the great way. The ideological experience of Chinese people is this kind of opposite and mutually complementary experience of two polarities. No matter what philosophical meaning you give to Tao, it is of no avail. Tao is a very profound life experience and aesthetic experience, but it has nothing to do with the proof system of truth.

From then on, true intelligence will not only be able to reason in the narrative framework, but also be able to speak the truth in the system of proof, not only be able to speak poetic wisdom in the perceptual experience of the world to satisfy emotional needs, but also be able to express logical analysis wisdom in the intellectual grasp of the world to meet the requirements of the scientific spirit, so intelligence can truly overcome the crisis of human nature.

5. Intelligence is not everything

Intelligence is only a tool to solve problems. If it is not combined with the customs and habits in daily life, the benevolence, justice, etiquette, wisdom, faith, and courage in ethics, and the statistical probability of boundary rules in the law, it is easy to become uncontrollable and overflow, causing disasters. True intelligence is not a panacea. It not only involves factual truth and falsehood issues, but also includes right and wrong issues of value, and it is closely related to the degree of responsibility. Therefore, in a strict sense, intelligence is a series of combined applications in many fields.

Mastering an amount of information and data does not mean getting closer to intelligence. For people, knowledge is a process rather than a static set of ideas. Sometimes it may make people feel desperate to discover the possibility of true laws with constructive efforts based on known facts. The fallacy will preempt people's minds if the framework of thinking is wrong. Therefore, the crux of the problem lies in the lack of a bridge between "immeasurable human subjective feelings" and "measurable objective material world." The resulting "real change" of human-machine- environment system interaction (it is precisely the development of a human cognitive calculation and suanji system based on deep situation awareness) firmly anchors the core issue of "human" It tries to describe people with unpredictable mathematical models instead of treating people as a lifeless atom. However, there are also problems with pure human factors. People who do not know how to do it or are stupid are huge hidden dangers and potential dangers in the loop.

For humans and machines, the solution of the big problem can be inferred when we divide the problem into several subproblems and solve the subproblems separately, but the dynamic programming of humans and the dynamic programming of machines are different; those experienced people can easily break a complex problem into small problems of different nature such as facts, values, and responsibilities to solve them. That is to say, they use different methods of facts, values, and responsibilities to make big things smaller and avoid being short-sighted, but the current machines are still far behind being able to resolve the problem of heterogeneity. Artificial intelligence will only compare (not analogy). Perhaps this is another bottleneck and difficulty of human intelligence: how does one effectively deal with heterogeneous informal issues?

The so-called human-machine intelligence has the ability and function to autonomously, actively, and appropriately handle the relationship between the subject and the object, and then recognize the trend, grasp the direction, and choose the path. There is not much intelligence in other control systems without human-like autonomy (adaptive).

Facts and values are relative, but the relative degree of the two is different. Modern artificial intelligence is always in contact with various fixed annotations and definitions, and these annotations or definitions are always inconsistent with themselves. Real intelligence is otherwise. Real intelligence is oriented to living objects (attribute annotations) and oriented to dynamic processes (relationship definitions). It not only involves "true" and "false" but it also includes "right" and "wrong" In the situation awareness of human natural intelligence, the theory of relativity also plays a role. State generates a new situation, and percept can generate a new sense (machines cannot do so), such as the shrinking effect of the general trend, which can produce attention, contempt, looking upward, looking downward, glaring, and facing up in vision after perception.

Each reference system in humans and machines has its own independent time. If the time of the two reference systems is not the same, they have the kind of relationship in corresponding situation awareness with first-order accuracy. Then two events that are considered to occur simultaneously in one frame of reference may be considered not to be simultaneous in the other frame of reference. Therefore, the inconsistency of information time, space, and intention is the key to human-machine hybrid intelligence.

I often hear someone say, "I believe that the laws of the universe should be simple and beautiful" but many people do not know that they need to look at it from a certain height to understand this simplicity and beauty. An oil painting is beautiful, but you may only see the spots in the oil painting if you get very, very close to it, which is neither simple nor beautiful. Similarly, if you want to understand and discover the laws of physics that are simpler and more beautiful, you must have a deeper understanding of the original theory and look at it from a higher height. And this kind of cognition, in-depth thinking about basic scientific issues, requires the participation of philosophy.

The gentleman focuses righteousness, while the villain focuses profit; human beings focus righteousness (right and wrong), and machines focus profit (gains and loss); human beings are a blend of affection and meaning, while machines are separated. Of course unreasonable people have to be excluded. Han Yu's (an ancient Chinese poet) definition of the word "righteousness" is "the appropriateness of action" means "appropriateness" which is equal to "should" But the traffic light problem still does not solve this problem of "righteousness" The blinking yellow light of the traffic light is an example of the rationale of the aforementioned human-machine hybrid interaction; for example, the blinking yellow light indicates differently for different people: "good people" stop, and "bad people" are still going through the crossing. Human rationality is evolved from sensibility, while machine rationality has not gone through this process, so it is impossible to simulate real human rationality or intelligence. In fact, most real human intelligence is instructive rather than prescriptive. Artificial intelligence is just the opposite.

The notion of causality, which constitutes an explanatory feature, can only be established in the case of a singular state of affairs explanation. But it does not hold or even is counterintuitive when this concept is extended to the interpretation of rules. Although the counterfactual theory of causal interpretation requires that the interpretation relationship must be an invariance relationship, the interpretation relationship cannot satisfy the invariance under intervention in some complex situations. According to Woodward's theory of the causal explanation, the explanatory relationship must have an invariant relationship. However, many biologic systems exhibit complex dynamics, including bifurcation, amplification, and phase changes. That is to say, in many cases, when we interpret the rules causally, the interpretation relationship does not satisfy the relationship of invariance under intervention. As J. Odenbaugh pointed out, in ecology, "It's almost impossible for us to manipulate the ecosystem in a systematic and controllable manner. Various factors are at work, and some of them just can be identified in a specific period." What is more, in many cases, many laws of cause and effect work together and cannot be changed individually. Causal explanation can explain the rules. Even though the existing causal explanation theory cannot explain all the rules, the causal explanation model will also explain more rules with the improvement of the causal explanation model.

6. The cornerstone of human intelligence may not really be mathematics

In Monet's view, the shape of an object is just a symbol of light, so he does not care about the specific shape but first observes and quickly records the reflected light and shadow during painting. With the stacking of brush-strokes and colors, the shape will naturally emerge. This original painting method is called "complementing shape with light" The "Sunrise Impression" may not be Monet's most outstanding work, but it touches the essence of impressionist art: **it uses a more direct and colorful way to express the various visual impressions of things without the pursuit of reality.** It tries to record the momentary feeling and that kind of hazy impression.

Doing research must be well-organized to form a sustainable ecologic development. Researchers can inherit the former research or find another way to do the research, but we must follow the trend. The state is the maximum possibility obtained with limited reality (time and space or state). It is often arranged and implemented in the planning and preparation stages. It can also be understood as a prelude to power. The combination of the two is the state. At present, the research context of strong artificial intelligence is not very clear, but the basic approach is still about data, algorithms, computing power, and experiments. It can be said that the cornerstone of artificial intelligence is mathematics. However, formal computers are always based on rational logical deductions, and human intelligence is often a combination of obscure logic. Therefore, some people predict that one of the subversive signs of strong intelligence in the future is probably whether a variety of integrative logical relations can be produced, and there should be various conflicts and contradictions between these complex logics from time to time, just like Monet's "Sunrise Impression" and Beethoven's "Symphony No. Five in C minor, Op. 67" (also known as "Fate Symphony"). Artificial intelligence is the empowering result of the integration of man and machine, in which empowerment is to empower functions rather than capabilities. Wittgenstein's transition from *tractatus* to philosophical studies is a transition from logic to illogical, a transition from function to ability, and a transition from weak intelligence to strong intelligence. At present, it is generally believed that the mechanism of human cognition is from state to form and then to situation and percept, from vision to feeling to perception and then to percept. Therefore, people have created artificial intelligence that can be used to defeat themselves in certain aspects (such as Go, etc.), and they feel panic from time to time, even feel depressed and pessimistic! As everyone knows, human beings have a more powerful ability that has not been given to artificial intelligence: that is the reverse process of "from state to form and then to situation and percept, from vision to feeling to perception and then to percept" Moreover, this feature can only be possessed by humans! The machine can calculate forward to form a number of solutions, from which people can choose, perform reverse interpretation computers, and finally make a decision based on experience.

Leibniz's thoughts on universal language and rational calculus should be the theoretical basis of Western artificial intelligence, which derived reference and meaning in Frege's philosophy of language, binary representation and set/logical operation in Boolean algebra, instruction coding and operation program in Turing machine, von Neumann structure, and so on. It is very possible that Hume's question (whether facts can deduce the value of the problem) is the strong AI ideological basis, whose nature can be seen as a form transformation problem with intentionality. The logic is different from value logic, for the former is relatively stable and does not vary from person to person, while the latter relatively changes and in some cases will be reversed of course. Hume's question also concerns the synthesis of calculation and suanji. Finally, we are repeating the most important thing: suanji may be a roundabout "irregular" hybrid deduction of many kinds of logic, if calculation is regarded as a relatively straight sequential deduction of logical rules. One of the hallmarks of future strong intelligence may be the ability to generate composite, merging logical relationships.

Mr. Xu Guozhi once introduced the origin of system theory. In the 1920s, Bell Telephone Company in the United States established Bell Laboratories, which was divided into Department of Parts and Department of Systems. In the late 1940s, people named the whole of some concepts, ideas, and methods introduced and created by the Bell Telephone Company in the expansion of the telephone network as "systems engineering" Since the middle of the 20th century, many scholars have used system names to name their research objects, such as computer integrating manufacturing system in control theory, management information system and decision support system in management science. With the advance of the times and the development of science and technology, people find that the interaction between things has become greater, and many problems have to be considered in general. So system science came into being. The system is like one of the basic concepts in mathematics: set. But different from a set, it is a heterogeneous set of classes. Different from traditional system science, a human-machine hybrid intelligent system is a complex system, which includes both scientific and nonscientific parts. Calculation is a quantitative isomorphic solution, while suanji is qualitative exotic reasoning. The difficulty of calculation combining with the suanji in human-machine environment system engineering lies in the implicit and indirect situation awareness. The difference between human and machine lies in the difference in the processing of state and trend. Human's suanji is better in state, while a machine's calculation is good at "situation" Situation refers to the state of objective fact, and "state" relates to the tendency of subjective value. Hilbert quoted Kant on the title page of the first edition of *The Foundations of Geometry*, "All human knowledge begins with intuition, thence goes to the concept, and ends with the idea." Perhaps this sentence is only half true! After all, in addition to dialectics, there are "change" and "evidence" The entanglement not only exists between quantum but also between the "state" and "situation" between "sense" and "percept" and between calculation and suanji.

Automation is a calculation system designed to perform repetitive tasks. An autonomous (independent) system refers to a system that performs tasks without human intervention. In the human-machine systems, we are particularly concerned with computer systems that perform complex reasoning tasks. But, can we also consider the autonomy of the human-machine hybrid, which means the calculation-suanji autonomous system derived from the human suanji + the machine's calculation.

Automation is driven by (deterministic) data computing without autonomous decision-making, while humans are driven by (dynamic) information and knowledge computing, which are able to handle unexpected situations and to try and verify. In a sense, HMI is a biophysical system that combines human's suanji intelligence with machine calculating automation. More importantly, human intelligence is aware that they are not intelligent, whereas machines are not. Humans can understand and use concepts outside the concept, while machines themselves do not have the ability and method to fit out reasonable concepts. They only grasp the tangible part of concepts and forget their intangible parts. Machines also sometimes mistake means for ends, and mistake results for causes. For example, in machine reinforcement learning, there are only gains and losses, and it is easy to form a "local optimal" and lose the "general trend" Just as the book "*Roots of Wisdom*" once said, "Doing good without realizing the benefits is like a wax gourd in grass." (If there is no reward in the process of doing good, it is like a wax gourd in the grass, it still thrives even though the human eye cannot see it.)

The key to interpretability lies in the trust generated by appropriate transparency. The key to trust lies in the agreement after understanding, in which understanding is the grasp of meaning, that is, the ability to organically integrate various potentially related things (facts, values, responsibilities, etc.). This world is composed of intelligence and non-intelligence. What can restrict the intelligent system in the future may include nonintelligent factors! It takes more than understanding to read Kant, which takes a great deal of imagination.

"One" can produce "many" and "many" can be condensed into "one" A long-term unity must be be followed by division, and a long-term division must be followed by unity. But, when do "unity" and "division" appear? How fast is the speed and acceleration of "division" and "unity" What role, besides reason, does emotion play in these transitions?

What is knowledge? What is a concept? What is understanding? What is consensus? Why does intelligence study philosophy, ethics, nonlogic? After the demand of artificial intelligence and applications emerges like bamboo shoots after spring rain, we still need to calm down and do in-depth thinking on the development of artificial intelligence. Thus we may get a deeper understanding of the intelligence and construct a mixing intelligence system of the artificial intelligence, intelligent, man-machine hybrid intelligent systems, which is able to meet people's expectation. It is hard for the old method of mending to achieve new breakthroughs. For example, people feel the same way about "knowledge" in the knowledge graph. Moreover, it is also said that when playing chess with the AI systems such as AlphaGo/AlphaZero, the best result is often not to "see the far view" within several steps, but to "look at the mouse eye" in front of the better.

Perhaps making an original work is not a technical work, but it is more like a subjective idea of the heart. It does not exist originally, but you have to convince people to believe in it. The challenge lies in a firm world view and insight before belief but not in engineering achieving methods.

CHAPTER 5
Military intelligence

Contents

1. The concept of military intelligence

As the top form of human intelligence, military intelligence is not military + artificial intelligence, but it includes both the machine's self-consistent process calculation and the human's directed suanji of

Integrated Human-Machine Intelligence
ISBN 978-0-323-99562-7
https://doi.org/10.1016/B978-0-323-99562-7.00005-X

contradiction. Military intelligence, which is like biologic evolution, does not pay attention to getting more powerful and more intelligent, but it focuses a lot on the appropriate adaptations of task execution. It is not a panacea curing all diseases, but a quasimedicine, whose highest state is to achieve the purpose of subduing people without fighting.

The current autonomy of military systems is equivalent to advanced automation under weak communication and no communication conditions, while modern military unmanned systems focus on mechanization + automation under statistical probability. Even if the equipment developed by science and technology is advanced, the product or system formed by it is only machine calculation. The mathematical basis of 0 and 1 remains unchanged, just like 5G and 6G. Without the appearance of intentionality and value, the system is still a machine in essence.

The essence of military intelligence is violent confrontation and competition, that is, to destroy the other side's game intention. The essence of artificial intelligence is service intelligence to meet the needs of the object. Military intelligence focuses on harming others, while civil intelligence delights in helping others. As the logic of calculation, AI is essentially a kind of "subject turning" The "strategy logic of military intelligence" takes human beings as the main body, studies the cognition, thinking, and intelligence of opponents, and emphasizes what should be done. Military intelligence not only involves means, but it also includes will and chance. The logic of AI calculation takes the calculation as the main body of information processing, focuses on what to do, and studies the processing mode of the calculation and the interaction between human and calculation.

The future of military intelligence is not a functional tool (hammer) but a capable piece of software + hardware + wetware, which is less about facts and forms and more about value and meaning. It will continue to transcend the pattern of services, industries, fields, and forward-looking strategic vision, and it is an important support for disruptive technologic innovation.

In the late 1950s, the consensus in the US military was that its command and control system could not meet the urgent needs of rapid decision-making in an increasingly complex and rapidly changing military environment. So in 1961, President Kennedy asked the military to improve its command and control system. The Department of Defense assigned Defense Advanced Research Projects Agency (DARPA) to run the program after major security concerns were raised. To that end, DARPA created the Office of Information Processing Technology and asked Joseph Licklider, a professor at the Massachusetts Institute of Technology, to be its first director. Although it is an urgent need of the military and an issue appointed by the president, DARPA does not get caught up in the immediate needs and specific problems of the services, but it carries out long-term and continuous research work based on the idea of "man-machine symbiosis" proposed by Licklider, considering man-machine interaction to be the essence of the command and control problem. Since then, the Information Processing Technology Office (IPTO) followed Licklider's ideas and gradually opened up many new fields in computer science and information processing technology and nurtured epoch-making and disruptive technologies such as ARPAnet, which has had a profound impact until today.

Military intelligence is neither unmanned nor autonomous. Autonomy refers to the degree to which one is in charge out of control of others, while unmanned refers to the degree to which all the scheduled operation tasks can be completed automatically under the condition of unmanned operation and assistance. On the contrary, military intelligence is mainly to realize higher-dimensional perception, insight, and implementation, carrying fraud and antifraud, which is the deep situation awareness of man-machine environment system integration, and it is the "steel" (equipment) + "qi" (spirit) of man-machine integration.

At present, many people think that military intelligence is military + AI, and some other people think that military intelligence is autonomous system or unmanned system. Most of them do not recognize the real nature of the military confrontation game. There is another problem of military intelligence that needs to be vigilant: the more sophisticated, accurate, and rapid computers made by simple machines are, the greater the danger will be. Because the enemy can hide the truth and show the false, create potential to deceive, and confuse the false with the true, the human-machine integration of military intelligence with the participation of experts becomes more important, urgent, and effective.

2. Development of military intelligence

The stages of the war situation are divided into the following parts.

2.1 Mechanical war and information war

The development stages of the forms of war have successively experienced the stage of mechanization, the stage of information technology, and the stage of intelligence. They are produced under different background conditions of different times, depending on the different material bases of the industrial age, the information age, and the intelligent age. The material basis of mechanization is mainly power equipment, petrochemical energy, and other physical entities and related technologies, while the material basis of information is mainly calculation and network hardware equipment and its operating software. An important precursor to intelligence is information technology, which is mainly based on massive data resources, parallel computing capabilities, and artificial intelligence algorithms provided by high-information technology.

Mechanization mainly improves the combat effectiveness of a single weapon by enhancing the power, firepower, and protective force of the weapon, and it improves the overall combat capability by means of intergenerational renewal and expansion of the numbers and scale of the weapon. Informatization is mainly to achieve information empowerment, network energy accumulation, system energy enhancement, software version upgrade, and system emergence to enhance the overall combat effectiveness through the construction of an information-based combat system, with information flow driving material flow and energy flow. On the other hand, intelligence is to give the combat system the ability of "learning" and "thinking" through artificial intelligence on the basis of high information, to improve the overall combat effectiveness in the way of rapid iterative evolution.

The mechanization mainly equips the army, which aims to improve the army's mobility, fire power, and protection, so the army can run and fly. The ultimate goal of mechanization is to make the weapons and equipment of all services and arms have more fierce firepower, faster speed, longer range, and stronger protection, and thus achieve the optimal mechanical performance indexes. The final standard of informatization is to make people or weapons equipped at the right time, the right place, in the right way to obtain and use the right information, thus to make the information acquisition, transmission, processing, sharing, security, and other performance indicators to be the best, and to achieve the actual battlefield transparency, command efficiency, strike precision, and intensive support. The pursuit goal of intelligence is to continuously improve the "INTELLIGENCE quotient" of single weapon equipment, the command information system, and even the whole combat system and simultaneously to improve its reliability, robustness, controllability, interpretability, and other related performance indicators.

2.2 Electronic cognitive war

The rapid growth of cyberspace is shaping a future world in which "everything is controlled by cyberspace" and giving birth to the national security law of "whoever controls cyberspace can control everything" At present, the world's major military powers are stepping up the planning of cyberspace national security strategy to seize the head start. A few countries are striving to seek military hegemony in cyberspace, forming cyber war forces, developing cyber attack weapons, developing cyber war regulations, and constantly strengthening cyber attack and deterrence capabilities.

Projects such as DARPA's Adaptive Radar Countermeasures, Behavioral Learning Adaptive Electronic War, and the Air Force Research Laboratory's Cognitive Electronic War Precise Reference Perception (PRESENCE) are examples of the development of this new cognitive electronic war technology. These cognitive electronic war (EW) technologies are expected to enable EW systems to lead new threat systems with wider frequency bands and greater radio frequency agility.

Cognitive electronic war technology has a broad application prospect, which will not only help to enhance the strength of electromagnetic countermeasures technology, but it also will have an important impact on information war and cyberspace war. Cognitive electronic war technology can realize autonomous scanning and positioning of the electromagnetic environment, determine the mode of electronic attack independently, and improve electromagnetic protection capability through strict spectrum control, which represents the development direction of future intelligent war.

Cognitive electronic war technology can effectively solve the problem of insufficient accuracy of situation awareness in traditional electronic war, avoiding the problem of exposing jamming signals and incurring anti-radiation strike due to high-power suppression means, which effectively improves the concealment and antidestruction of the electronic war system. The American Army's Urban Saber program is designed to improve spectrum control on the battlefield by using cognitive technologies to automatically detect, identify, classify, locate, and rapidly attack high-priority electronic war targets.

Cognitive electronic war technology will effectively adapt to the complex electromagnetic situation in the future battlefield, solving the problem of accurate situation awareness in the complex electromagnetic environment, and its real-time dynamic learning ability can respond quickly to the new complex environment. In the future, the cognitive electronic war, which integrates many high-tech technologies, will develop toward the direction of learning, thinking, reasoning, and memory.

2.3 Network-centric warfare

Network-centric warfare (NCW), usually called network-centric operations, is a new military guiding principle created by the US Department of Defense to turn the information advantage into a war advantage.

This is done by liaising separate but well-informed units on the ground through highly reliable networks to develop new organizational and combat methods. This network allows people to share more information, cooperation, and situation awareness, so it can make ministries coherent, command faster, and act more effectively in theory. This set of theories assumes that troops connected with a very reliable network update share information that will improve the quality of information and situation awareness. Situation awareness sharing allows cooperation and spontaneous collaboration, which significantly increases operations efficiency.

Through the networking of various combat units on the battlefield, the information advantage is changed into a combat advantage. The scattered troops can jointly perceive the battlefield situation and coordinate operations to give full play to the combat effectiveness of the combat style. Network-centric warfare (NCW) is a vital research achievement for the US military to promote the new Revolution of Military Affairs (RMA). It aims to improve information and command and control capabilities to enhance joint firepower and the ability needed to deal with targets. NCW is a war based on a new concept. It is fundamentally different from consumable warfare in the past. The speed of command operations and the self-synchronization between departments make it a fast and effective war.

The essence of NCW is to use computer information networks to carry out integrated command and control of troops or soldiers in various places. Its core is to use the network to share real-life information of all combat forces, grasp the battlefield situation in real-time, shorten decision-making time, and improve strike speed and accuracy. In the network-centered battle, commanders at all levels and even ordinary soldiers can use the network to exchange a large amount of visual information. Moreover, they can exchange views quickly, formulate combat plans, and solve various problems to carry out rapid, accurate, and continuous attacks on the enemy.

In the past, combat operations were mainly carried out around weapon platforms (such as tanks, warships, aircraft, etc.). Each platform obtained battlefield information by itself and then commanded the firepower system to carry out combat tasks in operations. The platform's mobility is conducive to implementing flexible and independent operations, limiting communication and sharing ability of information among platforms at the same time, thus affecting the overall combat effectiveness. It is precise because of the emergence of computer networks where it is possible to exchange information and share information among platforms, so battlefield sensors, command centers, and fire strike units can be formed into an organic unity to achieve actual joint operations. Therefore, this NCW that regards the network as the core and link can also be called network-based war. Therefore, the basic idea of NCW is to make full use of the network advantages of network platforms, obtain and consolidate their information advantages, and turn these information advantages into decision-making advantages. Compared with traditional method, NCW has three significant advantages:

1. fighting against common goals by gathering firepower simultaneously
2. improving troop protection through resources
3. forming a more effective and rapid "discovery → control" fighting sequence

NCW emphasizes the geographic decentralization of troops. In the past, due to limited capabilities, the adjustment of the army's combat strength had to be completed by repositioning, and the force could either get close to the enemy or the combat target to the greatest extent possible. As a result, the combat effectiveness of a decentralized deployment force could not be formed into a powerful fist, and it is impossible to quickly respond to the situation or concentrate on launching an assault because of the need for location adjustment and logistics. In contrast, information technology frees troops from the physical geographic location of the battlefield and enables them to move more effectively. Because of its clear grasp and understanding of the battlefield situation, the combat unit can gather firepower at any time instead of concentrating on attacking the enemy. In NCW firepower maneuver will completely replace the traditional force maneuver, so there is no longer a clear front line in the battle. The distinction between the front and back is not very obvious, and the strategic, operational, and tactical levels of war are increasingly diluted.

2.4 Algorithm warfare

Based on war intelligence, the US Department of Defense put forward the concept of "algorithm warfare" on April 26, 2017. And software that obtains much information from more information sources or algorithms that can replace manual data processing and provide people with data response suggestions are called "warfighting algorithms" At the same time, the US Department of Defense decided to set up a cross-functional group for algorithm warfare to promote the research of key technologies of "war algorithms" such as human intelligence, big data, and machine learning. This act of the US military seemed sudden, but actually the thought has been experienced a long time. It adapted to the urgent needs of modern warfare.

"War algorithm" originates from complex problems in the process of information warfare. With the expansion of the current battlefield in space, complex and diverse battlefield information sensors are spread all over land, sea, air, outer space, and electromagnetic network space. All kinds of intelligence reconnaissance, surveillance, and early warning information have exploded, and the resulting massive information data exceed the capabilities of intelligence analysts, which leads to severe problems such as the untimely collection of battlefield information, low timeliness of effectual information output, and feedback errors. At the same time, the introduction of new intelligent weapons and equipment such as drone swarms and swarm weapons and other new combat styles puts forward higher requirements for the timeliness, accuracy, and sensitivity of commander decision-making. Different data types and operations require standardized analysis algorithms to establish a data-independent analysis system, which can shorten the response time of observation, judgment, decision-making, and action loop, saving data bandwidth and effectively improving data processing and mining efficiency, thus reducing battlefield situation awareness. The uncertainty plays a role in crucial combat fields such as intelligent decision-making, command coordination, intelligence analysis, combat verification, and electromagnetic network attack and defense. As the war develops from physical and skill competition to intelligent competition, the artificial intelligence of war algorithms is linked to the command and control system and occupies a key position in it, which is the technical basis for realizing intelligent combat and building an intelligent army.

2.5 Mosaic warfare

The organization and planning of modern warfare will be cross-domain and cross-services. The US military has realized the importance of distributed, joint, and multidomain combat capabilities. Still, it will take years or even decades to develop and deploy related high-level networked architecture. To enable commanders to use the currently available systems to build the combat capabilities needed to win the war at combat speed, the Strategic Technology Office of the Defense Advanced Research Projects Agency put forward the concept of "mosaic warfare" in 2017. They sought to develop tools and procedures to reliably and flexibly connecting different systems, which combine a large number of low-cost sensors, command and control nodes, and weapon platforms and form a new asymmetric advantage against the enemy using networked combat to achieve the complexity at an efficient cost ratio.

The US military is currently developing more advanced fighters, submarines, and unmanned systems. However, with the global spread of military technology and high-tech systems, the strategic value of traditional technology platforms such as advanced satellites, stealth aircraft, or precision ammunition in the United States is declining. On the other hand, the fast update of the electronic component technology on the commercial market makes the new military system with a costly and decade-long development cycle outdated before delivery. Mosaic warfare aims to link together the simpler network systems to share information and fight collaboratively, among which consumption and information sharing ability is the key.

"Mosaic warfare" requires the combination of the system in different ways to achieve different effects. However, the existing weapons systems of the US military are not designed for the use of "mosaic warfare" They are more like jigsaw puzzles, which are carefully designed systems that can only play a specific component of a specific figure. The Strategic Technology Office aims to create technical architectures such as connectors, communication links, accurate navigation, and timing software to enable existing systems to work together.

"Mosaic warfare" can make the lethal chain more elastic (perception—decision—action decision-making). Since ancient times, The US military has optimized the "perception—decision—action" ring to be an "observe—orient—decide—act" (OODA) ring. Supposing the commander can separate the functions of the ring, various decision-makers can connect various sensor platforms to various operational platforms, thus bringing about the possibility of various arrangements and combinations, forcing the enemy to fight against various attack combinations. This makes the lethal chain more flexible. No matter what operations the enemy takes, the US military can always complete its lethal chain.

2.6 Multidomain warfare

The concept of "multidomain warfare" results from the theoretical exploration and research of the army and other services operations of the US Army over the past 10 years, focusing on the operational needs of armed conflicts with rival powers from 2025 to 2040, which is a brand-new battle concept driven by the "third offset strategy" As soon as this concept was released in October 2016, it was sought after by senior officials of the US Department of Defense, various services, operational commands, and research institutions. It became a hot topic of research in the US military and military research circles in 2017. From 2017 to early 2018, even if the leaders of the US state and the Department of Defense changed and the "third offset strategy" almost disappeared, the research and development and exploration of the US Army's "multidomain warfare" concept remained unabated, and many works were still progressing steadily.

With the continuous penetration and integration of new combat areas such as space, cyberspace, electromagnetic spectrum, and information environment into traditional combat areas such as land, sea, and air, joint operations will have a global combat space in the future. To coordinate and arrange possible combat operations from any corner of the world, "multidomain warfare" expands the original three-zone (rear, melee, depth) regional framework into a global framework of seven zones (battle strategy support area, campaign support area, tactical support area, melee zone, depth mobile zone, battle depth fire area, strategic deep fire zone).

The fundamental combat force envisaged in "multidomain warfare" is a flexible formation of multidomain hybrid, which requires allocating combat forces in the primary combat unit structure of land, sea, air, sky, network, and other fields, so the detachment can operate in multiple combat areas and release energy. Flexibility requires combat units to flexibly form relevant forces according to tasks to meet the changing operational needs. Such combat units must also respond quickly, reach the conflict area within a few days, and immediately start operations. They also should have strong survivability, taking the initiative and prudent action through task-based command according to the mission objectives when communication and navigation are blocked and communication with superiors is not smooth. They should have strong self-protection ability and the ability to carry out semi-independent operations without continuous supply and security flanks.

The winding mechanism of "multidomain warfare" can be expressed as forming an advantage window through cross-domain energy gathering, using the advantage window to promote the maneuverability of forces in various regions, promoting the development of the combat process toward favorable directions. This, in turn, creates more advantageous windows in linkage or concurrency. The combat process can be gradually promoted by creating and utilizing advantageous windows, ensuring that the coalition forces control the initiative while the opponents have to face many difficulties. Cross-domain energy gathering is the combat effectiveness of aggregating multiple domains of oneself. At a specific time and region, it acts on the opponent's specific combat domain to achieve the suppression of one or more enemy's combat domain capabilities. Cross-domain energy gathering is a new form of joint combat force integration. It has a low level of joint level, whose field is more comprehensive, the integration is more profound, and the accuracy is more acceptable.

The advantage window is a temporary advantage formed in a particular domain against the enemy and a weakness, error, or even system gap of the opponent. It may be manifested in the loss of the opponent's firepower, mechanical power, and protection in specific time and space, the loss of opponent's control of the network and electromagnetic space, the deviation of people's public opinion, or a wide gap formed by the concurrency of various domain effects of the opponents. The creation and utilization of the temporary advantage window reflect a deep understanding of the dynamic relationship between combat time, space, and purpose and the precise command and control of multiple forces and complex combat operations. It is a brand-new concept that transcends control power.

The theory of "multidomain warfare" divides the confrontation of strong players into three stages: competition, conflict, and return to competition. It is emphasized that the forward troop deployment will be continuously adjusted to developments during the competition stage, taking advantage of various opportunities to deploy troops to critical positions, breaking through the opponent's "anti-intervention or regional rejection" strategy, and changing the opponent's "rejection" area into the front-line area. Once the confrontation escalates to armed conflict, the network domain and space domain combat forces can immediately carry out operations, and the multidomain expeditionary combat forces can be sent to the theater within a few days to coordinate with the front-line deployment forces. Once the action is successful, the goal is achieved, and they return to competition based on ensuring their interests to the maximum extent, avoiding excessive stimulation of the opponents and causing the conflict to get out of control.

3. Development of US military intelligence

3.1 Three offset strategies

Since World War II, the United States has proposed three strategies with an "offset" nature. In the first, in the face of the financial crisis after the Korean War in 1953 and the threat of the Soviet Union, the United States proposed a "new look" strategy to offset the overwhelming conventional military strength advantage of the Soviet Union with nuclear technology advantages. However, with the improvement of Soviet nuclear capabilities and the formation of the Soviet-American nuclear balance, the first "offset strategy" lost its role and failed.

The second was in the mid- to late 1970s. In response to the postwar dilemma of Vietnam, especially the conventional military strength of the Soviet Union, the United States proposed an "offset strategy" led by precision strike technology with information technology as the core. Relying on its dominant position in technology and industry, the United States invested heavily in researching and developing new information technology to achieve "technology empowerment value" Through the use of satellite surveillance, global positioning, computer networks, accurate guidance, and other technologies, the combat effectiveness of existing weapons platforms dramatically improved, and the second "off-off strategy" was launched. At the same time, it promoted scientific and technologic innovation. The second "offset strategy" was considered to have successfully accelerated the Soviet Union's strategic recession and led to the collapse of the Soviet Union and the end of the Cold War.

The first offset was in the era of nuclear weapons and intercontinental ballistic missiles and satellite spies; the second was about stealth technology and precision guidance technology; the third is about applying technology, including autonomous learning systems. Military confrontation capabilities have long been upgraded to basic discipline research related to synthetic biology, quantum information science, cognitive neuroscience, human behavior modeling, and new engineering materials. Whoever has these cutting-edge technologies is likely to be in the leading position.

These three "offset suanji" are in the same line: they are long-term competitive strategies that seek to support and widen military advantages with technologic innovation against the background of the relative decline in national strength and the intensification of significant power challenges in the early stage of the war.

The purpose of the third "offset strategy" is to use advanced technologies such as artificial intelligence and autonomous capabilities to achieve a step-by-step leap in combat effectiveness, thus enhancing the conventional deterrence of the United States. Walker believes that the strategy includes technologic progress. Still, it is a concept of combat and organization based on regulations, training, and exercises, so the US military can use these technologies for combat and to gain advantages. The strategy is also related to the institutional strategy, organizing the Ministry of Defense to fight in the new dynamic environment.

The US Department of Defense stressed the need to attach importance to artificial intelligence and autonomy and integrate artificial intelligence and autonomy into combat networks, focusing on five aspects: autonomous learning systems for processing extensive data and judging paradigms, human-machine cooperation to achieve more timely relevant decisions, auxiliary combat through technical assistance (such as exoskeletons or wearable electronic equipment), advanced man-machine combat formations, such as joint combat of manned and unmanned systems, and network-enabling weapons and high-speed weapons, such as directional capabilities, electromagnetic rail guns, and hypersonic weapons.

3.2 Four stages of development experienced by the US military DARPA military intelligence

After World War II, electronic and computer technology made rapid progress, laying the foundation for replacing human tasks with machines. In the early 1960s, DARPA (it was called ARPA at that time) began to intervene in autonomous technology research and soon became the primary research institution in this field. DARPA realizes that artificial intelligence can meet a large number of national security needs. In artificial intelligence projects, the automation of intelligence-related capabilities is promoted through the integration of achievements in computer science, mathematics, probability, statistics, and cognitive science. The scope of research has gradually moved from speech recognition, language translation to extensive data analysis, intelligence analysis, genome and medicine, vision and robotics, unmanned driving and navigation, and other fields.

Although DARPA has been developing autonomous technology for a long time, its research projects related to autonomy have not been carried out in a fixed technical field but scattered in many different fields. It was not until 2014 that autonomous technology was officially demarcated under the Office of Defense Science.

Autonomous technology covers various fields such as communication, command and control, data processing, etc. To be focused, this book classifies independent technology according to the projects studied by DARPA's new autonomous technology territory, combined with the classification of DARPA technology in reports/books such as DARPA "Technical Achievements" (1990) and "Strategic Computing" (2002). The scope of research is limited to autonomous technologies related to land, sea, and air robots and autonomous technologies related to intelligent assists. According to the time node, DARPA's research on autonomous technology can be roughly divided into four stages.

3.2.1 Artificial intelligence research stage

The development of artificial intelligence in the United States is mainly due to the support of DARPA. In the early 1960s, DARPA developed time-sharing operation techniques of computers in the MAC program and began its initial research on artificial intelligence technology. However, it was not until the late 1960s that artificial intelligence was included in the DARPA budget as a separate research project. By the mid-1970s, DARPA had become the prominent supporter of artificial intelligence research in the United States and promoted the practical application of artificial intelligence technology, such as automatic speech recognition and image resolution. In the late 1970s, artificial intelligence was widely used and in some military systems. In 1983, artificial intelligence technology became a vital component of the DARPA strategic accounting project.

In the research of artificial intelligence, DARPA not only supports basic research, such as knowledge expression, problem-solving, and natural language structure, but it also supports applied research, such as applied research in the fields of expert systems, automatic programming, robotics, and computer vision.

3.2.2 Strategic computing project stage

In the 1980s, international research on computer systems (especially in Japan) increased, and DARPA felt that its dominant position in the field of computing was threatened. Therefore, in 1983, DARPA established the Strategic Computing Project to improve the advantages of all computing and information services. AI has become an essential part of strategic computing projects.

Some research projects in AI made significant progress before entering the Strategic Computing Project, while others faced major technical problems that are unsustainable. Therefore, strategic computing projects still invest in all technology fields in AI projects but focus more on technologies that can continue to achieve progress. The four projects that have attracted attention are the voice recognition project, which can support navigation assistance and combat management, natural language development, which is the basis of combat management, visual technology, which is the foundation of autonomous drones, and expert systems that can be used in all applications.

3.2.3 Development stage from 1994 to 2014

After the Strategic Computing Project, the Advanced Technology Office (ATO) and later the IPTO continued to have research on relevant autonomous technologies, and dozens of technology researches have been carried out over 20 years, including ATO's tactical maneuver robot man project (mainly using remote control technology), ITO's maneuver autonomous robot software project and software of distributed robot project, MTO's distributed robot project, etc.

3.2.4 Stage of establishing the field of ownership (2014 to present)

In the second quarter of 2014, DARPA's Defense Science Office established a new research domain: autonomy (semiautonomy). It mainly studies hardware and computing tools to enable the system to work correctly in an environment lacking infrastructure through intermittent connection alone. At present, research projects in this field include autonomous robot manipulation (ARM), Fast lightweight autonomy (FLA) project, and Mixed Initiative Control of Automata-teams (MICA) project.

3.3 The main areas of the development of US DARPA military intelligence

3.3.1 Speech recognition

The original project was the Speech Understanding Research Program, launched in the early 1970s. In this program, DARPA supported multiple research institutions to use different methods to conduct speech recognition research. The best results are CMU's Hearsay-II technology and BBN's HWIM (Hear What I Mean) technology. Among them, Hearsay-II puts forward the forward-looking concept of using parallel asynchronous processes to piecemeal process people's speech content. At the same time, BBN's HWIM improves the accuracy of vocabulary recognition by decoding complex speech logic rules.

In the 1980s, DARPA began to use statistical methods to study voice recognition technology and developed a series of voice recognition systems such as Sphinx, BYBLOS, DECIPHER, etc., which have been able to recognize speech continuously throughout sentences.

After 2000, DARPA began to develop a system for human-machine interaction through dialogue, which can also learn from experience in dialogue with different people and provide personalized services. In 2001, DARPA developed a translation device for individual soldiers. After the 9/11 attack, voice recognition technology received further attention. A translation device called Phraselator that can perform one-way translation was launched.

In 2005, DARPA launched the Global Autonomous Language Exploitation project, which seeks technologies that can translate printed materials, web pages, news, and television broadcasts in standard Arabic and Chinese in real time. It is planned to achieve 95% accuracy of text file translation and 90% voice file translation in 2010.

3.3.2 Environmental perception technology

Environmental perception technology mainly involves the identification and application of various sensor information. DARPA's originally aims to develop a technology that automatically or semiautomatically analyses military photos and related pictures. With the deepening of research, especially the development of unmanned systems (mainly unmanned vehicles), and the strict requirements for information input, DARPA's projects have gradually moved from identifying static information to the direction of sensing and recognizing dynamic information being developed.

In 1976, DARPA began the image recognition (IU) project, whose original goal was to develop a technique that automatically or semiautomatically analyses military photos and related pictures in 5 years. Project participants include Massachusetts Institute of Technology, Stanford University, University of Rochester, SRI, and Honeywell. In 1979, the goal of the project was expanded, and graphic drawing technology was added. By 1981, the projects expected to be completed within 5 years were not terminated but continued into 2001.

In the fiscal year 2001, DARPA launched the PerceptOR project to solve the problem of environmental perception, whose purpose is to develop a new unmanned vehicle perception system that requires that the system is agile enough to ensure that unmanned vehicles can carry out tasks in off-road environments and perform tasks in various battlefield environments and under different weather conditions. In 2005, the project completed the phased study and then transferred it to the "Unattended Combat System Ground Drone Vehicle Integration Product" project for system development and testing.

In 2010, DARPA launched the "Heart and Eye" project. The purpose of the project is to develop an intelligent vision system. Through visual input alone, people can learn general applications and reproduce them through action.

3.3.3 Technology of artificial intelligence

DARPA began working on technology of artificial intelligence in the 1970s, when the IPTO supported research at Stanford and MIT (such as the machine-assisted cognition project later), but artificial intelligence (including robotics) was not DARPA's focus point at that moment.

In the early 1980s, DARPA stepped up research on autonomous air carriers and autonomous ground vehicles and autonomous maritime vehicles (which were later known as robot killers), but the research did not achieve its intended goal, and the relevant research results provided the basis for later strategic computing projects.

In 1985, DARPA's study on artificial intelligence (including killer robots) became a part of the Smart Weapons Program under the responsibility of the Tactical Technology Office.

In 1999, under the computer and communication project, the development of intelligent system and software technology was set up to develop an artificial intelligence system that can actively and autonomously provide all kinds of auxiliary information to soldiers.

In 2006, DARPA began its Integrated Learning Program, which aims to combine domain knowledge and common sense to create a reasoning and deduction system that can learn like a human and can be used for a variety of complex tasks. Such a system would significantly expand the types of tasks in computer learning and lay the foundation for the development of automatic systems for complex tasks.

In 2010, DARPA began funding deep learning projects with the goal of building a general-purpose machine learning engine. Deep learning can accomplish artificial intelligence tasks that require highly abstract features, such as speech recognition, image recognition and retrieval, natural language understanding, and so on. The deep model is an artificial neural network with multiple hidden layers. The multilayer nonlinear structure makes it have strong features expression and complex task modeling ability. At present, deep learning is the intelligent learning method closest to the human brain, which brings artificial intelligence to a new level and will have a far-reaching impact on a large number of products and services.

Deep learning originates from the research of artificial neural networks. The model used for deep learning is deep neural networks, that is, neural networks containing multiple hidden layers. Deep learning utilizes the hidden layer in the model and makes the original input be transformed layer by layer into shallow features, middle features, high features, and finally the task objective through the way of feature combination.

3.3.4 Technology of robot autonomous control

The research on the control technology of military robots (including unmanned vehicles) can be traced back to the 1930s, when it mainly focused on the remote-control technology of wheeled/tracked vehicles. Later, the control technology gradually developed from remote control to semi-autonomous and autonomous control, from the control of the movement of wheeled/tracked vehicles to the control of the movement of biped and multilegged robots. And at the same time, it also added the control technology of functional components (such as manipulator) that can complete complex operations.

In the1980s, DARPA invested development of wheeled vehicle control technology in the development of ground unmanned vehicles and funded vehicle control technology research with the Defense Department's Joint Robotics Program after the 1990s. Since the beginning of 21st century, DARPA has successively launched the Unmanned Vehicle Challenge, which triggered the research on vehicle control technology in a wider range.

In the1980s, DARPA developed unmanned ground vehicles, and DARPA developed a small ground robot (vehicle) with the support of the Future Combat System program in 2001. The robot developed in this project adopts the motion mode of walking or crawling, forming a new control scheme.

In 2008, DARPA proposed the learning mobility project. The projects aim to develop a new generation of learning algorithms to enable unmanned robots to successfully pass through large and irregular obstacles. More importantly, through the continuous accumulation of experience, these algorithms will enable the robot to independently learn to overcome the actual terrain that is more complex than that set by human coding.

The project is a collaborative and competitive effort by six research teams, where each team is provided the same small quadruped robot (Little Dog) manufactured by the Boston Institute of dynamics. To reduce the complexity of remote sensing in navigation issues, each team also provides a motion capture system designed by Vicon. In this way, the research team can focus on finding the best algorithm to solve the problem and judge the rugged terrain changes under the same hardware conditions.

In February 2010, DARPA set up ARM, a new robotic autonomous technology. The goal of the project is to develop a controller with high autonomy and suitable for multiple military tasks, so the robot can perform human level tasks quickly and at the least cost. In the absence of human control, ARM makes full use of its own vision, power, and touch sensors to flexibly master and complete 18 different tasks, and the commercial components of ARM robot include arm, neck, head sensors, and so on. The current robot control system can protect life and reduce casualties, but its ability is limited in a multitask environment, which requires more human intervention and takes a long time to complete the task.

3.3.5 Autonomous marshaling and coordination technology

Before the 1990s, a single functional combined structure was developed, and the information circulation structure was simplified and single. From 1991 to 1996, some emerging technologies emerged, such as hierarchic link state routing protocol Isis, operating system kernel mach, and subcontracting communication and switching technology myrinet, which is widely used in interconnected clusters such as workstations, PCs, servers, blade servers, or single board computers. DARPA establishes a distributed layout that is a fixed control center connecting many of the same terminals to realize the differentiation of individual organizational units.

From 1997 to 2001, to solve the problems caused by the increase in the number and types of terminals, coordinate the dynamic allocation of resources among various parts, provide more services effectively, quickly, and accurately, and realize the feedback mechanism of real-time communication and dynamic planning among various departments, DARPA later designed an integrated computing device with more complex relationship connection, DARPA/SC21Concept (2010), which met the cooperation requirements of aircraft combat formations and ship combat formations to deal with complex war tasks through the coordination at planning and different mission levels.

4. Key factors to win the intelligent war in the future

4.1 Human-machine environment system integration

In recent years, the alpha series, an outstanding representative of AI, have made brilliant achievements in chess and other games, but it is still a kind of relevance machine learning and reasoning under closed conditions, while the root of the military intelligent game is still the learning and understanding of people with mixed causality and relevance in an open environment. This kind of learning can produce a certain range of uncertain tacit knowledge and order rules (as children learn), and this understanding can relate those things that seem to be unrelated. There are signs that the war of the future may be one of the integration of human-machine and environmental systems.

Sun Tzu said, "Knowing both your enemy and your friends, you will win a hundred battles without danger." Here, "knowing" includes both human perception and machine perception. The difference between human perception and machine perception is that humans can have intentionality without formalization, while a machine's understanding of intention is not as flexible and deep as humans. The "enemy" here includes the opponent as well as the equipment and environment. The "friends" here also includes one's own people, machine, and environment. So, there will be no intelligence without people and no future war without artificial intelligence. Real intelligence, or artificial intelligence, is not something that can be achieved by abstract mathematical systems. Mathematics is just a tool that can only achieve functions rather than ability: only humans can produce real ability. So the artificial intelligence is a product of the interaction of a human, machine, and environmental system. The war of the future is also the result of the calculation of machine combining human suanji. It is a kind of suanji combining calculation, or a kind of insight. In fact, simple calculation can be faster, more accurate, more nimble, more dangerous, and more gullible. As the old Chinese saying goes, "Clever people may be victims of their own cleverness" A famous Chinese idiom,

"Blessing hides in disguise," shows that calculation is not as good as a man's suanji and insight.

Recently, services of the United States have proposed multidomain warfare, whole-domain warfare, mosaic warfare, and other modes for future combat modes. It is a kind of human-machine environment system engineering, and it is the dispersion and aggregation of human, machine, and environment. It is also the comprehension, mixing, and fusion of both the distributed phenomenal calculation of many symbols and phenomenological representation suanji of many nonsymbols, and it is also mutual complement of mechanics, information, knowledge, experience, artificial intelligence, and intelligence.

Therefore, cracking the intelligent mechanism of a human-machine hybrid will become the key to victory in the future war. Any division of labor is limited by scale and scope. Function allocation in human-machine hybrid intelligence is one kind of division of labor, and capacity allocation is another kind. Function allocation is passive and caused by external demand, while capability allocation is active and internally driven. Human or humanoid directional preprocessing is very important in the complex, heterogeneous, nonstructured, and nonlinear data/information/knowledge. When the problem domain is initially narrowed, the machine's advantages of boundness, rapidness, and accuracy can play their roles. In addition, when a large amount of data/information/knowledge is obtained, the machine can initially map them into a few areas, and then humans can further process and analyze it. The assimilation, adaptation, and crossing balance of these two processes are roughly the organic hybrid of man and machine.

4.2 Intelligent coordinated operations

The future war is not only an intelligent war, but also a war of wisdom. The future war should not only break the boundary of formal mathematical calculation, but it should also break the boundary of logic suanji of traditional thinking. It is a new calculation-suanji game system combining the complementary advantages of human, machine, and environment. It is a bit like education, where the school's job is to teach the knowledge to the students (a bit like machine learning), but education is not just about teaching the knowledge, but it is also about discovering the logic behind the knowledge, or something deeper. For example, when we are teaching calculation, we really need to think about what is behind the calculation.

We should first develop the students' sense of numbers, then teach them the concept of calculation, what is plus and what is minus, and then teach them how to apply it, thus developing insight.

In the intelligent war, cooperative warfare is a necessary means. Given the continued spread and proliferation of nuclear weapons, the cost of future wars between nations (no matter if they are large or small) will become ever higher. In a sense, they are both partners, competitors, and strategic rivals (we should not only prevent nuclear, biochemical, and intelligent weapons from getting out of control, but also destroy each other's will and defeat the opponent). If men are regarded as strength and women as wisdom, then the future war should be a feminine war, at least it will be a hybrid war of men and women.

No matter how artificial intelligence develops, the future belongs to human beings. Human beings should jointly define the game rules of future wars and decide the fate of artificial intelligence, rather than the fate of human beings being determined by artificial intelligence. The reason lies in that artificial intelligence is logical, while future wars are not only logical but also have a large number of nonlogical factors. In the face of the enemy's strong electromagnetic spectrum and cyberspace combat capability, the seamless integration and coordination of various systems such as information communication, command and control systems, intelligence, surveillance, and reconnaissance will also be a major test when the various services carry out multidomain operations.

Therefore, the future war is the effective combination of human, machine, and environment and the cooperation of multiple fields, forming an intelligent cooperative combat mode.

5. Challenges of deep situation awareness in intelligent warfare

5.1 The problem of the human-machine hybrid

On the surface, the military intelligence in various countries develops rapidly. In fact, there is a fatal defect in the process of military intelligence in various countries, and the reason lies in the failure to deeply deal with the intelligent problem of the human-machine hybrid. Any subversive scientific and technologic progress can be traced back to the understanding of basic concepts. For example, all human behaviors have a purpose, and this purpose is value. Purposefulness can be divided into far, medium, and near purpose, and its value degree is respectively large, medium, and small.

In addition to valuing causal reasoning, human beings are more powerful than artificial intelligence in various variable characteristics, variable representation, variable understanding, variable judgment, variable prediction, and variable execution. Strictly speaking, the current artificial intelligence technology application scenario is very narrow, in the early stage of the computing intelligence and cognitive intelligence, which does not take the initiative to accurately portray scene and situations, and the greatest difficulty in intelligent science is effectively characterizing scene or context. The idea of military intelligence in the past and modern times is to train a bunch of artificial intelligence algorithms and bind their own military application scenarios.

In general, this artificial intelligence technology is to formalize causal reasoning and data calculation with symbols, behavior, connectionism, objective facts, rarely involving perceptual factors, causality judgment, or decision-making. The "depth" in the deep situation awareness refers to the integration of fact and value. State and trend refer to objective factual data and information, and an objective part of knowledge (such as saliency, temporal and spatial parameters, etc.), which is simply called the fact chain, but sense, percept, and other parameters of the subjective value are called the value chain. The deep situation awareness is that the fact chain and value chain intertwined in a "double helix" structure, which can realize effective judgment and accurate decision-making functions. In addition, the "human" focuses on the subjective value to control suanji, while the "machine" focuses on the objective fact to process calculation, which jointly forms a "double helix" structure. How to achieve a proper match between two "double helix" structures (space—time, salience, expectation, effort, value, etc.) is a difficult problem that has not been solved by all countries. In a sense, deep situation awareness not only solves the prominence of time and space contradictions in a human-machine environment system, but also the selectivity of fact contradictions, value contradictions, and responsibility contradictions. Conflict is competition, and decision-making involves risk.

The advantage of human-machine hybrid intelligence lies in the full integration of the advantages of man and machine. However, human beings are accustomed to scenario-oriented and flexible knowledge expression and multifactor weighing and reflective reasoning decision-making, which are very different from the machine's data input, axiomatic reasoning, logical decision-making mechanism. Once it is not possible to organically integrate the human and machine, it will reduce the efficiency of intelligent decision-making system of the human-machine hybrid when they interact with each other. At present, there are still many theoretical problems to be solved in knowledge representation and the decision mechanism of the human-machine hybrid.

The main problem of human-machine hybrid knowledge representation is the lack of elastic knowledge base that can fuse sensor data with a commander's knowledge and be adapted to actual combat scenarios. Human commanders have complete military theoretical knowledge, such as "science of tactics" "weapon science" and "topography" and they have specific representation habits for organizational preparation, determination, firepower preparation, and real-time combat operations. Therefore, if the machine wants to understand the semantic expression of the commander in a specific task scenario, it needs to automatically analyze the task, enemy situation, tactics, terrain, and other factors to form a comprehensive situation judgment. Instead of enumerating all factors in advance based on the traditional "programming thinking", we should "perceive, understand, and learn" the battlefield situation to make the knowledge base flexible and metabolized and solve the problem of the consistency of signifier and the signified of human-machine tactical knowledge.

The main problem of the human-machine hybrid decision-making mechanism is the lack of personalized intelligent decision-making mechanism based on man—machine communication. The commander's styles are different, and the intelligent system that can realize efficient human—machine cooperation must be a personalized intelligent system. A "Personalized" intelligent system is not a simple adaptation and accommodation of the machine to the commander's habit. It should establish a kind of human—machine communication framework and mechanism. The decision-making suggestions of the system may be a supplement to the commander's thinking, or they may be completely contrary to the commander's command style. Feedback can be obtained through continuous practice, and the human-machine hybrid decision-making ability can be developed iteratively. Finally, personalized auxiliary decision-making system can be realized to achieve the optimal matching between man and machine.

5.2 Uncertainty on the battlefield

Clausewitz, a famous military theorist, believes that war is a fog with a great deal of uncertainty and unknowability. The unknowable here is unpredictable and cannot be judged. From the development trend of modern artificial intelligence, there are many hidden dangers of the human-machine hybrid in wars in the foreseeable future that are still unsolved, as follows.

(1) In a complex game environment, humans and machines absorb and use limited information in a specific time and space. For humans, the greater the pressure is, the more misunderstood information there is, and the more likely it leads to perplexity, confusion, and accidents. For machines, it is still very difficult for a computer to learn, understand, and predict cross-domain, unstructured data.

(2) The wide distribution of information needed for decision-making in war in time, space, and emotion determines that it is still difficult to obtain some key information in specific situations, and to coordinate and integrate the important objective physical data collected by machines with the subjective information and knowledge acquired by humans.

(3) A large number of nonlinear characteristics and unexpected variability in future wars often lead to unpredictability in the process and of operation results, and formalized logic reasoning based on axioms has been far from meeting the needs of complex and changeable decision-making in the war.

5.3 The problem of humankind

"Cross-domain collaboration" is a "problem of people" The methods that "multidomain warfare" solves for the "multidomain collaboration" problem can be summarized in two terms. The first is "convergence" which refers to "the integration of capabilities across domains, environments, and functions to achieve a purpose in time and physical space." The second is "integration of systems" which focuses not only on the people and processes needed to achieve "multidomain collaboration" but also on technical solutions. So far, "multidomain collaboration" has not recognized that current systems and programming projects are independent of each other as "chimneys" and multidomain maneuvers and multidomain firepower require a "human's" solutions. As automation, machine learning, artificial intelligence, and other technologies mature, US military adversaries will seek to apply these technologic capabilities to further challenge the United States. It is the responsibility of the US military to break out of the existing "chimney" scheme and design a new one with a human-machine formation behind it, as Walker requested.

On May 12th, 2020, Peter Hickman, a US defense expert, published an article titled "The Key to Success in Future wars is People" The article argues that AI will make a significant contribution to the evolution of war as the nature of warfare continues to evolve, but there is still a risk of overestimating the pace of technologic change and the role advanced technology will play in future victories. An overemphasis on technology will cause competitors to discover blind spots and exploit them. There is nothing wrong with pursuing cutting-edge technology, but the key to success in future wars will still be people.

The rapid development and wide application of artificial intelligence has become a leading factor in the new round of scientific and technologic revolution and industrial revolution, and a core force to promote the innovation of weapons and equipment, the process of military revolution, and the change of war patterns. Using the machine to assist the commander to complete the command and decision task, man and machine will inevitably form a dependent relationship once the auxiliary comes into being. In the future, intelligent warfare will be cognition-centered warfare, with intelligence being the dominant force. Intelligence will weigh more than firepower and mobility, and the pursuit will be to control energy with intelligence. Deep situation awareness in the human-machine hybrid penetrates the situation understanding, decision-making, accusations, and other links, and it will play a role of multiplication, transcending, and dynamics in each link.

The shortcoming of science is that it denies the reality of individuation that cannot be controlled and repeated. There are bound to be flaws based on this science. Man, especially every individual, is a natural, uncontrollable, and unrepeatable subject of individuation. You cannot say that he does not exist. From this point of view, the essence of the human-machine hybrid is to help science perfect its shortcomings and limitations.

The advantage of big data is the general repeatability of controlled experiments, so common laws can be found: try to locate following the map. However, this is also a disadvantage of big data. It is easy to ignore the emergence of new things, the nonrepeatable part of controlled experiments, showing the effect of "carve on gunwale of a moving boat to seek a falling sword" It is also true that some controlled experiments cannot be repeated, but this is not within the scope of science. It used to be "blind men touching an elephant to guess what an elephant looks like," while it becomes "human-machine searching for a sword" now.

In the future battlefield, the combat confrontation situation is highly complex and fast changing, and a variety of information intersections form huge amounts of data. Thus it is difficult to achieve rapid and accurate processing only by human brains, but a commander (human + machine) can deal with the rapidly changing battlefield and complete command and control task only when humans and machines integrate based on the database, and other technology groups like the internet. With the increase of autonomous capability of unmanned autonomous systems and the enhancement of artificial intelligence cluster function, autonomous decision-making gradually appears. Once the command system realizes the Intellectualization of different functions, the time of perception, understanding, and prediction will be greatly reduced, and the efficiency will be significantly improved. With the additional help of pattern recognition for battlefield sensor image processing and an optimal algorithm for combat decision-making, the command system will be endowed with more advanced and perfect decision-making ability and gradually realize the joint operation between human and machine.

CHAPTER 6

Research on the application of antiartificial intelligence in the military field

Contents

Artificial intelligence is characterized by the integration of technical, social, legal, ethical, and military attributes. On the one hand, it has facilitated enormous technologic and social changes and profoundly influenced military weaponry. Artificial intelligence has become a national strategy and a new core competitiveness. On the other hand, it may bring many risks and challenges, and the problem that AI may lose control and cause harm cannot be ignored.

At present, any military power in the world regards anti-AI as the most important military technology in the future, thus increasing investment in anti-AI weapons. The US Department of Defense has begun work on a master plan to build systems to deploy anti-AI to the military. The use of anti-AI in the military field is booming, and the widespread use of anti-AI devices will not only revolutionize traditional forms of warfare but also have a significant impact on military leadership and control theory. Research on anti-AI applications in the military should be accelerated, and during research on anti-AI weapons and equipment, we must also ensure that we update and improve command and control principles under the anti-AI warfare conditions.

Integrated Human-Machine Intelligence
ISBN 978-0-323-99562-7
https://doi.org/10.1016/B978-0-323-99562-7.00006-1

1. Antiartificial intelligence in the military field

Artificial intelligence is a new technical science that researches and develops theories, methods, technologies, and application systems for simulating, extending, and expanding human intelligence. Antiartificial intelligence is the theory, method, and technology that can counter the opponent's artificial intelligence algorithm and equipment from the angles of data, algorithm, and hardware under the condition of human-machine cooperation. Anti-AI includes disabling the artificial intelligence of the opponents, misleading their artificial intelligence, obtaining their real intention of the artificial intelligence, and even carrying out counterattack.

The nature of AI and anti-AI is a game. The use of anti-AI in the military field is giving rise to a whole new form of warfare. Although the application of antiartificial intelligence is still regarded as the reality in the future in the military field, the application of antiartificial intelligence in the military field has begun to take shape in the local wars in the new century with the development of antiartificial intelligence. Some anti-AI military weapons have been used in actual combat, fundamentally changing the mode and methods of modern warfare.

2. The nature of antiartificial intelligence

The essence of antiartificial intelligence is fraud and antifraud. Sun Tzu's strategy of war goes like this, "There can never be too much deception in war" and "Those who have plans will win more easily than those who do not." Instead of trying to detect all fraud through anti-AI, we should learn to distinguish the real from the fake and move on through the fog of fraud.

Anti-AI is in its infancy. So far, the development of anti-AI technology has gone through two stages. Initially, most anti-AI technologies work by misleading or confusing machine learning models or training data, which is a simple and crude method. However, since machine learning models are usually trained in a closed environment, it is difficult to get external interference. With the development of neural networks, an adversarial neural network opens the second technical route of antiartificial intelligence technology. Researchers can use neural networks around adversarial data to generate feedback that causes machine learning models to make false judgments during recognition and action, which is similar to machine learning techniques that can achieve primary anti-AI effects.

Gaming has always been an important research topic in the field of antiartificial intelligence. According to whether the game information can be fully understood, it can be divided into complete and incomplete information sets. In the complete information set, all the information of the game can be completely obtained by all the players in the game. For example, in go and chess, both sides can fully understand all the fragmentary information and the opponent's action plan. An incomplete information set is a set in which participants do not have access to complete information, and only part of the information is visible. In mahjong or poker games, for example, a player has no control over another player's distribution or hand and can only make an optimal decision based on the situation at hand.

At present, anti-AI practices in the military field are developing rapidly, but there are also many crises. At the moment, anti-AI can only do some basic work, and it does not understand why it should do so in most cases. It just follows what the results of the data processing tell it to do to get an optimal solution. If an anti-AI system does not fully understand its function or surroundings, it can have dangerous results. Especially in military war, anything can happen, which can not be completely solved or predicted by data alone. At this time, antiartificial intelligence may perform wrong actions and cause incalculable consequences. A slight deviation in the training data of anti-AI in the military field may have hidden security risks. If the attacker uses malicious data copy training mode, it will lead to a major error in military implementation of anti-AI.

3. The necessity of antiartificial intelligence research

From the point of view of military defense, it is necessary to study anti-AI technology. AI has injected instability into the independent deployment of and uncontrolled factors of all countries. Decisions made at machine speed rather than human speed also add to the crisis in the machine decision-making. In ongoing conflicts, the parties use intelligent autonomous weapons to gain a military advantage at the outset, which provides them with a powerful fighting force in battle. It is quite difficult to define artificial intelligence and its rights. Even if the strategic system is stable, it may launch attacks to avoid threats, increasing the probability of accidental attacks in peaceful interactions. With intelligent autonomous weapons, the risk factor in war preparation increases. Intelligent autonomous weapons may go out of control and launch attacks intentionally or unintentionally.

The use of AI in the military has increased military uncertainty and made countries feel that their security is threatened.

In war, there are facts and values, and between facts and values, there is a possibility that may be space, time, or a unit. Fixed targets in war do not change much over time, and their space does not change much, but their value does. Mosaics, for example, are about finding objects, time, or space that these events focus on. AlphaGo can only talk about things, but it cannot read between the lines and so on.

The likelihood of an adversary invading is low because there is no crisis or because there is no accidental military action. The country's highest decision-makers must believe that there is a threat to one or more fundamental values. Preparation seems to be the beginning of the threat, and then general deterrence kicks in. Strong deterrence works when a war is brewing or taking place between two countries. During the Cold War, the common understanding of deterrence was almost entirely nuclear. The United States wishes above all to deter nuclear aggression, and the threat of nuclear weapons is the last resort it uses to deter such aggression. However, the United States has also sought to deter all significant aggression by other adversaries, and it has developed strong conventional military forces to support its conventional deterrence. Today, as during the Cold War, the United States has a strategy of deterrence designed to deter aggression against American territory and against American allies in Europe and East Asia. The use of anti-AI in the military could lead to unexplained conflicts, and inappropriate autonomous action may lead to unexpected escalation. The existence of intelligent systems will bring chances of technical accidents and failures, especially when executors do not have the ability to guarantee security, so accidents and false positives will in turn affect decision-making. In addition, though anti-AI upgrades only for defense but not for conflict and attack, it could escalate into smart rivalries between great powers once they are seen as upgrades by other countries. Therefore, to prevent accidental attack by AI weapons, it is necessary to accelerate the research of anti-AI application in the military field.

4. Antiartificial intelligence development in the US military

The US military has not yet explicitly proposed the concept of anti-AI technology, but a number of programs it has established are in the direction of anti-AI application in the military field. It is hoped that its

ARTIFICIAL intelligence and autonomous model can be continuously optimized to defend and even counter the artificial intelligence technology of its opponents. At present, it has conformed the true intention of its opponents should be analyzed from the "gray area" and the true purpose of its opponents should be analyzed from the insight of the situation. In the future, the investment in interpretability and prevention of strategic deception will be increased. So, we need to be proactive and push for anti-AI applications in the military.

In 2017, DARPA launched the Machine Lifelong Learning Project (M2S) to explore the application of analogical learning methods in ARTIFICIAL intelligence, seeking new breakthroughs in the next generation of ARTIFICIAL intelligence, enabling it to learn on the spot and improve its performance. In the real world, it does not need checking or networking. The ability to pursue independent learning enables the system to adapt to new situations without prior programming and training.

In August 2017, DARPA's "mosaic" concept was put forward, which is not limited to system designs and interoperability standards of any specific organization, service, or enterprise, but it focuses on the development of reliably connecting the application of various nodes and tools, seeking a different system of fast, smart, strategic combination and decomposition, realizing the infinite variety of operational effectiveness.

In 2018, DARPA released a program called "Compass" to help combatants understand an opponent's real purpose by quantifying various attacks. This program solved the problem in two ways. First, it is used to determine the adversary's actions and objectives, and then to determine whether the plan is working properly, such as location, timing, and movement. But before we could understand them, we needed to turn the data we acquired into information through artificial intelligence, and understand the different meanings of information and knowledge. This was the beginning of game theory. Game theory was then incorporated into the AI technology to determine the most effective action based on the opponent's real intentions scheme.

At the beginning of 2018, the Knowledge-directed Artificial Intelligence Reasoning Over Schemas (KAIROS) artificial intelligence project was officially launched. The US military expects to use the KAIROS project to improve location percept, early warning, intelligence procedures, and war intelligence capabilities. Specifically, in normal coordination mode, every country plans and implements hidden strategic steps. During war, the military forces of different countries adopt different strategies. The

KAIROS project hopes to build a system that can obtain the "intelligence behind the intelligence" which is expected to have stronger monitoring and warning, intelligence processes, and intelligent decision-making functions.

On the basis of the above plan, the "Active Interpretation of Different Sources" (AIDA) project will explore the control of key multisource fuzzy information data sources, develop a "dynamic engine" and generate data for actual data. Obtained from various sources, clearly explained events, situations, and trends are added, having complex quantification, so this program means solving potential conflicts and deceptions in the fog of war. For fraudulent data and hostile attacks, simulated data and public war data will create a test site to evaluate the risks of machine learning. At the same time, it will focus on upgrading antijamming user machine learning algorithms and integrating them into the prototype system. To prevent the enemy from interfering with our artificial intelligence, it will check the execution process of artificial intelligence through interpretable artificial intelligence, ensure the correctness of execution and achieving the effect of antiartificial intelligence.

5. Antiartificial intelligence and deep situation awareness

The meaning of deep situation awareness refers to "the perception of situation awareness, which is a kind of human-machine intelligence including both human intelligence and machine intelligence (artificial intelligence)." It is the mixing of signifier + signified, not only involving the attributes of things (signifier, feeling), but also relating to the relationship between them (significant, perception) and understanding not only the surface but also the deep meaning under the surface. It is based on Endsley's understanding of the subject situation (including use for input, processing, and output). It is a global analysis of system trends, including human, machines (objects), environment (nature, society), and their relationships. There are two feedback mechanisms, including domestic and global quantitative forecasts and evaluations, including self-organization, self-adaptation, another organization, and mutual adaptation, which are both autonomous systems waiting to be selected and autonomous systems-predictive selection-monitoring the effects of adjustment information.

Most of the data-driven artificial intelligence can be attributed to an optimization problem. For example, supervised classification discrimination learning is to make the classifier obtain a certain minimum error rate on the representative training data. Generally, we will assume that the training data can properly reflect the overall distribution; otherwise the generalization ability of the trained classifier is very questionable. However, in practice, people rarely test whether this hypothesis is true, especially in the case of high-dimensional samples, because the data is relatively sparsely distributed in space, so hypothesis testing is difficult to achieve, and the "curse of dimensionality" is commonplace.

We do not know how the population is distributed, the mechanism of data generation, and whether the observation samples are "qualified" to represent the population. Under this premise, even if there are a large number of samples to train the learning machine, bias will inevitably occur. Therefore, pure data-driven machine learning always contains certain risks. In particular, when we have some prior knowledge of the data generation mechanism but are limited by machine learning methods and cannot be expressed, our concerns about the lack of interpretability and potential of the model will further intensify. No matter whether the model is good or bad, we do not know the reason behind it, and we can evaluate the generalization ability and robustness of the model. Therefore, human thinking should be above data, and especially the understanding of causality (not only factual causality, but more importantly, value causality) should precede data expression.

The recent progress in more in-depth research is mainly due to the increase in computing power. For example, deep learning is the continuation and deepening of artificial neural networks, and its computing power pushes data-based algorithms to a higher level. People think that data can answer all incomprehensible questions and can be proved by intelligent data mining technology.

Data are very important, but they cannot be used as the only cause for decision-making (a lot of data will also play a disturbing role). These "cause" models with knowledge or experience are essential in helping robots transition from artificial intelligence to the application of artificial intelligence in the military field. Huge data analysis and data-based methods are only available in civilian forecasts. The application of antiartificial intelligence in the military requires intervention and illogical actions, so the machine can make decisions more in line with expectations. "Intervention frees people and machines from active exploration of passive observation and resorting to causal reasoning" to expand the imagination space, thereby overcoming the fog of the real world.

On the basis of causal reasoning, antiartificial intelligence conducts in-depth situation awareness of the battlefield, which is not only the acquisition and processing of information. Military antiartificial intelligence must also be able to get the true purpose of the other party in the information. In the "gray area" where artificial intelligence cannot handle the problem, antiartificial intelligence can use prior knowledge and situation awareness to make the best solution.

6. Antiartificial intelligence and the human-machine hybrid

The rapid development of human-machine hybrid is defined as "human-machine system engineering" that is "human-machine," which is a system that studies the best match between humans, equipment, and environmental systems, involving integration, performance, and management and feedback. The research goal of the system overall design is the human-machine environment, optimization and visibility, safety, robustness, harmony, and effective coordination of the entire system.

The key to intelligent systems is to be used "just right" and the key to human intelligence is to actively prevalue it to "just right" advance. The key to human-machine hybrid intelligence is to "appropriately" organize "active arrangements" and "passive use" sequence. There are not only suanji but also computers in computers combining with suanji, which can traverse nonfamily similarities. Calculation is to use the acquired data to calculate unknown data, and suanji is a purposeful estimation. The calculation starts with conditions, and the suanji starts with noncondition. All computers must use in-scope consensus rules for reasoning, but suanji are used to imagine nonscope and nonconsensus rules. The "calculation" of calculation is reasoning, while the "calculation" of suanji is imagination. The "calculation" of calculation is a known thing, and the "calculation" of suanji is the unknown. Data is an important point of natural interaction between humans and computers. British scholar Tim Jordan pointed out, "The massive amount of information makes it impossible to use this information effectively. If specific information cannot be found when the following two situations happens: there is some information that cannot be absorbed; the information is poorly organized."

The future antiartificial intelligence system is at least an intelligent system of autonomous integration of human-machine environmental systems. Calculation means that the flow of information includes input, processing, output, and feedback. One of the main development goals of antiartificial intelligence is human-machine hybrid intelligence. At present, strong artificial intelligence, human-like artificial intelligence, and general intelligence are far away from us. It is the integrated intelligence of humans and machines that carries the implementation of intelligence. Human-machine hybrid intelligence will learn how to achieve the best human-machine cooperation. As far as the design and optimization of human-machine environment systems are concerned, the ability to identify calculation and machine designs is very complex. Usually, it involves two basic issues, one of which is the fusion of human intentions and machine intentions. The so-called intention is the direction of consciousness. It is difficult for a machine to handle anything that can be changed and reversed, or those opposite things, but machines can expand the storage space at will, easily calculate, and can be formalized and symbolized. When and how do humans and machines intervene and react to each other? Under multiple constraints, time and accuracy have become very important. Therefore, how to fully integrate machine computing power and human brain cognition is a very important core issue in human-machine integrated intelligence.

At the same time, military antiartificial intelligence has another key point: the more precise and accurate the calculation is, the more likely it is to be used by the enemy. The enemy deceives by concealing the truth and revealing the false. So the hybrid of human and machine is very important, because the hybrid of human and machine intelligence is a complex field, not a single subject. The links of antiartificial intelligence can include input, processing, output, feedback, synthesis, and so on. In the input link, we need to split, merge, and exchange data, information, and knowledge, so the other party cannot obtain useful information, and at the same time, it can mislead the other party. In the processing link, it is necessary to block information processing, make its internal processing nonaxiom and axiom divergence, and make it at a loss for information processing. In the output link, make opponent's intuition and logical decision deviate, make them distrust the decision, so as to interfere with the opponent's final decision. In the feedback link, it is necessary to make the reflection and feedback paradoxical, so that it is confused or even rejected by the feedback information, so it cannot further absorb the previous case information. In the integration stage, it is necessary to make the opponent's situation awareness contradictory and make it impossible to understand and even contradict each other with higher-level information in the information aggregation and synthesis stage, to achieve the effect of defeating others without fighting.

To solve the problem of integrating humans and machines into military situation awareness, we must first break the inertia of different senses and break the traditional time relationship, including maps, perceptions, knowledge maps, and state maps. For humans, machines are tools for self-development and part of self-percept. We can understand our own mistakes through the advantages of machines, understand our abilities through machine errors, and then compensate or improve each other. Due to the lack of dualism, human-machine hybrid mixing is not yet recognized by most people. Nowadays, more and more human-machine hybrid interactions are continuously optimized. Although this is not satisfactory and there are still gaps, what is worth looking forward to in the future is that people will also discover themselves while manufacturing machines.

At present, the development of antiartificial intelligence and human-machine hybrid is still in its infancy. The first and most important problem of integrating antiartificial intelligence and human-machine hybrid interaction is how to integrate the antiartificial intelligence function of the machine with human-machine intelligence. In the application stage, the distribution of human-machine forces in the human-machine mixture is obvious, so no effective synergy can be produced. Humans continue to expand their cognitive abilities in the learning they are doing, so they can better understand the changing conditions in a complex environment. Because of the ability of association, people can create the ability of cross-domain integration, but cognitive ability runs counter to artificial intelligence thinking. The way to activate human-like thinking ability is to achieve a breakthrough in the integration of antiartificial intelligence and human-machine. Giulio Tononi's comprehensive information theory points out that intelligent systems must obtain information quickly. At the same time, the development of machines that can perform cognitive processing requires collective consciousness between humans and machines. Therefore, a fast and effective two-way information interaction must be established between humans and machines. The basis of two-way information interaction is abstract information. For computers, it has the ability to abstractly define the restrictive environment of matter. The more abstract the representation is, the more it can adapt to different situations. At the same time, high-level intangible abilities will also be transformed into universal migratory abilities, thus surpassing the limits of human thought.

7. Methods and strategies for antiartificial intelligence

By evaluating the military potential of each country's antiartificial intelligence plan and the system level of antiartificial intelligence implementation solutions provided by each country, all the impact is evaluated in the design, development, testing, or using the process, mainly including training data, algorithms, and system management. These data are introduced after testing to monitor on-site behavior and integrate the system with other human-machine hybrid interaction processes to control attacks. To evaluate the new combat thinking of each country, we should understand the situation at the system level and also understand the decision-making process based on antiartificial intelligence, how to control decision-making, and whether the use of antiartificial intelligence and autonomous combat concepts will cause errors and self-upgrade at any time.

Based on other combat or strategic war scenarios, antiartificial intelligence war simulation is a particularly effective tool to better understand intelligent warfare. A simulated war in a certain geographic area is conducted to test combat capabilities. The more scenarios and the more opponents and allies there are, the more possible it is for the inference to lead to different results.

In-depth study of the opponent's autonomous system and its use of the autonomous system is not only to understand one's own system, but also to understand the opponent's abilities. It is important to understand antiartificial intelligence and how to use autonomous systems and concepts to improve interaction with other countries or regions, so planners can better predict opponents' decisions.

The controllers of different artificial intelligence attacks are struggling to find opportunities to obtain the highest return. We can increase the cost of their attacks and reduce the benefits of their successful attacks to weaken the attackers' interest in them. As the organization's cyber security plan matures, their attack value will be reduced.

Task automation and malicious mass attacks further reduce the barrier safety factor, making it easier for attackers to enter and perform operations. Therefore, antiartificial intelligence can focus on defense and reducing the amount of attack. In a military war full of challenges, antiartificial intelligence technology needs to shorten the attacker's ability to remain anonymous and the distance from the victim, thereby reducing the difficulty of antireconnaissance. As a defender, antiartificial intelligence must be 100% successful in preventing the attack, and the attacker only needs to succeed once. Organizations must focus on cultivating the right capabilities and build a team to use processes and technology to reduce this asymmetry.

Although antiartificial intelligence and autonomy are reducing variability, cost, increasing scale, and controlling errors, attackers can also use artificial intelligence to break the balance and gain an advantage. Attackers can automatically manipulate the most resource-intensive elements in the attack process, while avoiding control barriers deployed against them. Therefore, we need antiartificial intelligence to scan for vulnerabilities quickly, find and remedy vulnerabilities faster than attackers, and prevent attackers from using this as a breakthrough point to concentrate their efforts on attacks.

A simple countermeasure to the risks of artificial intelligence attacks and changes in the threat landscape is to implement a high-pressure security culture. The defense team can adopt a risk-based approach to determine governance processes and substantive thresholds, so that defense leaders can know their cyber security situation and propose reasonable measures for continuous improvement. Using antiartificial intelligence and other technologies to improve the safe operation of operations and technical teams can obtain more logistical support. For example, we can realize the autonomy of resource or time-intensive processes through antiartificial intelligence, greatly reducing the time required to complete conventional safety processes. For the defense team, the increased efficiency of the safety process reduces the friction that is likely to occur in subsequent safety regulations. The development of antiartificial intelligence technology will bring more opportunities to improve war security and maintain a balance between risks and rewards.

8. Suggestions

The mechanization, automation, and informatization in military affairs have changed the "situation" and "sense" while anti-AI used in the army may change the "state" and "percept" as well as the "knowledge" of "boundary percept" related to the war in the future. Compared to the traditional "situation awareness" it is much deeper and comprehensive, forming a profound situation awareness. Human-machine coordination is a form of military intelligence, which integrates with the environmental system of human-machine, featuring a faster, more cooperative, more unsafe, and more threatening with less autonomy character. Thus it has to become much clearer within more conditions and limitations. Mutual agreement has to be timely and effective. Evaluating the inherent limitation of military situation awareness and the external limitation of human-machine, the main conclusion is that people have to cope with the intelligent weapon in the rival whatever the strategies and tactics. In the countless military applications, humans and machines have to be in the system at the same time.

The application of anti-AI weapons, on the one hand, has great advantages in precision strikes, reducing manpower costs, increasing operational flexibility, and preventing terrorist attacks. On the other hand, it faces challenges and threats, such as undermining international humanitarianism and triggering an arms race. From a global perspective, the solution of this problem requires all countries to work together.

At the same time, we must pay attention to improving the standards of military anti-AI algorithms. The military application of anti-AI technology is different from social applications. When anti-AI is used in the civilian fields, the training data required is very rich, and the scenarios are also relatively fixed, so the related algorithms can function well, but in the military field, especially in actual combat, due to the complexity, antagonism, and the variability of the battlefield environment, it is quite hard to get the training data required for anti-AI systems, and the effect of related algorithms will be compromised. This is a problem that anti-AI technology must face and solve in the process of military application. Meanwhile, the formulation of relevant military standards must keep up with the military application of anti-AI technology to ensure that it can meet the functionality, interoperability, and security requirements of the military field and optimize the effect of anti-AI in the military field.

CHAPTER 7

Deep situation awareness

Contents

It is safe to say that human civilization is a process where humans constantly recognize the world and themselves. If we intend to trace back the origin of AI, we have to understand human history and approaches of human knowledge acquisition. It is known that cognition is the process of collecting and filtering, processing, predicting output, and adjusting feedback focusing on the useful data: information. From the earliest human civilization of Mesopotamia (more than 6000 years ago), ancient Egyptian civilization (about 6000 years ago), to the Greek culture (about 3000 years ago), considered to be the origin of modern Western civilization, the essence reflects the relationship between human and object or, more exactly, objective object, which is the cultural basis for the rapid development of science and technology. The ancient Indian civilization contains beliefs between human and god. The later ancient Chinese civilization, still continuing among the four ancient civilizations, focuses on communication between individuals as well as human and environment, which may be an important reason why Chinese civilization continues till now.

Throughout the process of system interaction between human, machines (objects), and environment, the generation, circulation, processing, mutation, bending, amplification, attenuation, and disappearance of cognitive data are always there. How do we maintain all the possible stability and continuity in the variable process? People have proposed numerous theories and models and utilized many tools and methods to find effective answers and universal laws from nature and society. The "heliocentric theory" of Copernicus in the 16th century transferred authority of religion to science gradually. From then on, experiments and logic have reconstructed a completely different world of time and space over the centuries, reducing physiologic load, mental load, even mental load.

Integrated Human-Machine Intelligence
ISBN 978-0-323-99562-7
https://doi.org/10.1016/B978-0-323-99562-7.00007-3

With the evolution of scientific thought, considerable progress has been made in technology. System theory, cybernetics, and information theory have not faded, and dissipative structure theory, synergy theory, and mutation theory have been on the stage. Electronic tubes, transistors, and integrated circuits have not yet disappeared. Nano, supercalculations, and quantum communication technologies are trying to play. The AI ideas and technologies, born in the 1940s, are frontier directions based on these fields. However, due to ambiguity of cognitive mechanism, the lack of mathematical modeling, and the limitations of computing hardware, AI is unable to grow rapidly. Judging from the current research progress in mathematics and hardware, it will be difficult to make breakthroughs in a short time. Therefore, how to make a breakthrough from the cognitive mechanism has become the initial choice of many scientists. This chapter aims to provide a preliminary introduction and review of in-depth situation awareness, to promote research and application of the theory in China.

1. Deep situation awareness: the origin of the theory

In June 2013, Mica R. Endsley, a female scientist known for her research on situation awareness, was appointed officially as the new chief scientist of the US Air Force by the US Department of the Air Force. She graduated from the University of Southern California in 1990, majoring in industrial and systems engineering. She, as well as her predecessor Mark T. Maybury, focused on cognitive engineering in human-machine hybrid interaction, which changed the convention that chief scientists of the US Air Force major in aerospace or electro-mechanical engineering before September 2010. Such appointment of chief scientists with a professional background in cognitive science is also quite popular among other branches of the US military, which means the manufacturing and processing fields dominated by hardware institutions will not be as popular as the command and control system with the theme of software intelligence in military and civilian technology development trends in the future.

Coincidentally, while AI, automation and other majors around the world are seriously studying situation awareness technology, the global calculation community is working hard to analyze the context awareness (CA) algorithm. The linguisticians are also very enthusiastic in grammar, semantics, pragmatics, and other aspects in natural language processing. Situation awareness in psychology also is growing as a hotspot at the moment. Analytical philosophy is the mainstream of Western philosophy as a philosophical school, whose methods can be roughly divided into two types: the analysis method of artificial language and the analysis method of everyday language. Of course, the current research of major branches of cognitive science, such as cognitive neuroscience, also focuses on brain consciousness, trying to figure out the process of human consciousness through the structure and working pattern of the human brain.

Now people live in a human-machine-environment (natural, social) system with increasingly active information. Consequently, the command and control system regulates the ongoing process through the interaction among human, machine, and environment and the input, processing, output, and feedback of information, thereby reducing or eliminating the uncertain result.

As for the core of the command and control system, Mica R. Endsley proposed a dynamic decision-making situation awareness model. The specific structure is shown in Fig. 7.1.

Situation awareness in this model can be divided into three levels. Each stage is necessary but not sufficient to precede the next stage. The model follows an information processing chain, from perception through interpretation to predictive planning, from low level to high level. The first level is the perception of each component in the environment, that is, the input of information. The second level is a comprehensive understanding of the current situation, that is, the processing of information. The third level is the prediction and planning of the subsequent situation, that is, the output of information.

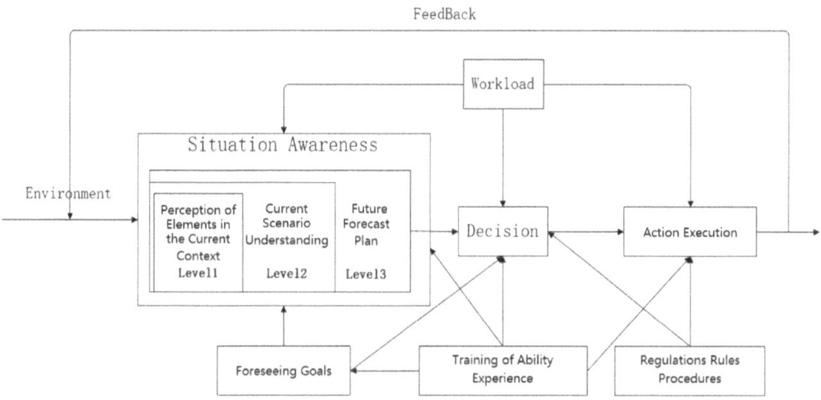

Figure 7.1 Dynamic decision situation awareness model (Endsley, 2000).

- Generally speaking, human, machine, environment (nature, society), and other components that constitute a specific situation often undergo rapid changes. In this fast-paced situation evolution, with an absence of a comprehensive perception and understanding of the situation due to limited time and information, the accurate quantitative prediction of the future situation may be greatly compromised, but it should not affect the qualitative analysis of the future situation. In the era of big data, with clarification of logical relationships between each component and its interference components, such as repulsion, attraction, competition, and risk-taking, it is crucial for AI systems to establish qualitative and quantitative comprehensive decision-making models based on discrete programming and continuous probability, even on emotions, insights, and reflection on objective situation.

In short, big data mining, without a deep understanding of relationship between data representations (especially heterogeneous variant data) is unreliable, based on which the intelligent prediction system cannot be reliable either.

In addition, in an intelligent prediction system, it is difficult to distinguish management defects from technical failures. How does one conceptualize nonconceptual problems? How do we homogenize heterogeneous problems? How can we integrate unreliable components into a reliable system? How does one minimize human errors and maximize the effectiveness of machine and environment through the front/back (rigid, flexible) feedback systems of the intelligent prediction system? H.A. Simon, Turing Award winner in 1975 and Nobel Prize winner in 1978, proposed a clever strategy: limited rationality, that is, to extend the nonconceptual and unstructured components in an infinite range into flexible concepts and structures that can be manipulated in limited time and space. In this way, the nonlinear and uncertain system can be linearized and processed satisfactorily. We are not seeking to find a needle in the haystack, but only satisfied with finding a needle in a bowl of water. At last the seemingly irrelevant things are related together, so intelligent prediction becomes smarter.

However, in practical engineering applications, due to various factors (subjective and objective) and imperfect processing methods, the current situation awareness theory and technology still have many shortcomings.

The establishment and maintenance of situation awareness can be a difficult process for people in different fields and environments. It is reported that pilots spend most of their time ensuring an accurate psychologic description of what happened in real time. The same is true for many other fields where the system is complex and must handle large amounts of real-time information and that information is rapidly changing and difficult to obtain.

2. Decryption of deep situation awareness

The deep situation awareness is "the perception of situation awareness, which is a kind of human-machine intelligence, combining both human intelligence and machine intelligence." It refers to property (referring, feeling), and it is related to the relationship between them (referred, perception). It can understand the original meaning, but it also understand the overtones. It is a kind of expectation-selection-prediction-control system with information correction with autonomous and automatic aggregation effects and compensation, featured with both self-organization, self-adaptation, and other organization, mutual adaptation, both partial quantitative calculation and prediction and global qualitative calculation and evaluation. It is based on Ensley's situation awareness (including information input, processing, and output links), coupled with the overall system trend analysis of human and machine (things into the environment (nature, society)) and their interrelationships, resulting in a "soft/hard" adjustment feedback mechanism. If the diffuse reflection of objects forms vision, deep situation awareness is as if turning on a switch in a dark room to obtain the origin of an object.

Deep situation awareness is to fully organize the system in a specific environment to complete the theme task using a comprehensive manifestation of various human cognitive activities, such as purpose, feeling, attention, motivation, prediction, automatism, motor skills, planning, pattern recognition, decision-making, motivation, experience and knowledge extraction, storage, execution, feedback, etc. It can operate in the context of insufficient information and resources, and it can also function in the context of information and resource overload.

Through experimental simulation and on-site investigation and analysis, it is believed that there is a "leapfrog" phenomenon (automatic response) in the deep situation awareness system, that is, directly entering the output control phase from the information input phase (skipping the information processing integration phase), which is mainly due to the clear task theme, the concentration of organization/individual attention, and the long-term targeted training for a conditional response. Just like someone walks while chewing gum, chatting, and opening an umbrella, they can unconsciously coordinate the order of various natural activities. So does the system. The system performs almost according to perfect automatic control, rather than a conscious response to regular conditions. This is in line with the saying in the book *The Quest of Consciousness*: in the process of learning, conscious participation will affect efficiency. Compared with ordinary situation awareness systems, such information sampling will be more discrete, especially in the information filtering after perceiving various stimuli, showing a strong ability of "removal of falsehood and storage of truth". The basic function of this information "filter" is to allow the designated signal to pass through relatively smoothly, while attenuating other signals. To highlight useful signals, it can suppress and reduce interference and noise signals for a higher signal-to-noise ratio or selection of purpose. For each stimulus object, it includes not only useful information characteristics but also other characteristics, and the deep situation awareness system can accurately grasp the key information characteristics of the stimulus object (it can be understood as the ability to "see the spot and know the leopard"), so it can form the cognitive ability of step-type artificial intelligence for a quick search, compare and refine and optimize the pruning planning and prediction of operations research, and can perform thematic tasks automatically and quickly. For ordinary situation awareness systems, for the cognitive response capabilities of deep situation awareness systems are not formed, the perceived stimulus objects include not only useful information features but also some other features, so it is massive in the amount of information sampling but slow in information fusion and forecast planning and weak in execution.

Time- and task-pressured, "experienced" deep situation awareness systems are often based on discrete empirical thinking schema/script cognitive decision-making activities (rather than evaluation based). These schema/script cognitive activities form the basis for an automatic model (that is, not requiring analysis at every step). They are based on previous experience accumulation to react and act, rather than through conventional statistical probability methods for decision-making. (The situational assessment in basic cognitive decision-making is based on scheme and script. Scheme is a type of concept or description of event, which is the basis for forming long-term memory organization. In the process of top-down information control processing, the information of a perceived event can be mapped according to the best matching existential thinking schema. However, the automatic processing of bottom-up information, according to the mental schema aroused by the perceived event, is adjusted due to inconsistent matching, and it matches the latest changed mental schema structure through active search.)

On the other hand, sometimes the deep situation awareness system is forced to make conscious analysis and decision-making on some changed task situations (the automatic mode can no longer guarantee the accuracy requirements of accurate operation), but the deep situation awareness system rarely shifts its attention to distracted nonsubject or background factors. This phenomenon may be related to complex training rules. The general situation awareness system is required to operate according to a specific program, which is triggered by threshold of cognition (that is, when the specified information is stimulated). In fact, the dynamic situation often changes the threshold. And the deep situation awareness system has formed a kind of internal experience through numerous practices and training experience. The implicit dynamic triggers the threshold value of situational cognition. That is, it is activated when it encounters the key information feature that is useful rather than being prescribed.

The extraction of information from a top-down process relies on (is at least affected by) previous knowledge of the characteristics of objects, while the extraction of information from a bottom-up process relies on current stimulus. Therefore, any process involving recognition is a top-down process, that is, the organization process of known information. The top-down process has been confirmed to have an impact on depth perception and optical illusion. Top-down and bottom-up processes can be processed simultaneously.

In most normal situations, the situation awareness system achieves the goal in the top-down process, while in abnormal or emergency situations, the situation awareness system may reach a new goal in the bottom-up process. In any case, the deep situation awareness system should be proactive in the context (such as using a feed-forward control strategy to stay ahead of changes) rather than reactive (such as using a feedback control strategy to keep up with the changes), which is very important. This proactive strategy can be obtained by training in response to abnormal situations or emergencies.

Under the real complex background, comprehensive research on the deep situation awareness system and technology has been carried out. According to the information transmission mechanism in the process of the human-machine-environment system, the construction of accurate and reliable mathematical models has become the goal for researchers. Human cognition experience shows that people have the ability to search for specific targets in complex environments and select and process specific target information. This search and selection process is called focus attention. In the cases of multibatch, multitarget, and multitask, it is a big problem to obtain the required information quickly and effectively. How does one apply the environmental focus and self-focus mechanisms of the human cognitive system to the learning of the multimodule deep situation awareness technology system, and how does one determine the input of the attention mechanism according to the processing task, so with the entire deep situation awareness under the control of an attention mechanism, the system can complete information processing tasks effectively and form efficient and accurate information output, which may provide a new way to solve the above problems. The first problem to be solved is how to build a moderately large-scale multimodule deep situation awareness technology system. In addition, how to control the integration and coordination of the various functional modules of the system is an important issue to be resolved.

Through research, the technical problems of deep situation awareness are as follows. First, the deep situation awareness process is an active behavior instead of a passive response to the environment. The deep situation awareness system is stimulated by environmental information by collection, filtration, change of situation analysis strategy, extraction of invariance from the dynamic information flow, and production of near-perceptual operation or control under the interaction of the human-machine hybrid environment. Second, the calculation in the deep situation awareness technology is dynamic and nonlinear (similar to cognitive technology calculation). It is usually not necessary to calculate all the problems clearly but to calculate the required information. Moreover, the calculation in the deep situation awareness technology should be adaptive, while the command and control system should be self-adaptive. The characteristics should change with the interaction with the outside world. Therefore, the calculation in deep situation awareness technology should result from the interaction of the external environment, equipment, and human cognitive sensors, and the three are indispensable.

Research on deep situation awareness system technology is based on human behavior characteristics, that is, research on the perception and response capabilities of organizations in uncertain dynamic environments. It plays a key role in emergency command and organization systems for major events in the social system (wars, natural disasters, financial crises, etc.), rapid processing of accidents in complex industrial systems, system reconstruction and repair, and design and management of humanoid robots in complex environments.

3. The construction of meaning

In the deep situation awareness system, we aim at construction of the meaning framework of the situation, not construction of situation, and then to realize in-depth prediction and planning in many uncertain situations.

Sense sometimes means fragments. Knowledge is the establishment of associations (relationships) at the same time. Human perception and knowing processes are often performed at the same time (other than the machine), and people can perform physically, psychologically, and physiologically at the same time. The sense and knowledge of attributes and relationships can also be mixed with feelings and knowledge. For a long time, a certain kind of intuition or emotion will be generated, from irrelevant to weak, from weak to relevant, from relevant to strong and even a "leapfrog phenomena". Analogy plays a very important role in this process, bridging between tacit knowledge with explicit rules/probabilities. According to phenomenology, the key to consciousness is perception, which is the ability to perceive surrounding objects and the world. The perception of object is the action that one can do with the object obtained from the integration of the interactive experience of oneself and the object. For example, the perception of an apple on a nearby table is that it can be eaten. It can be held in the hand when one walks over. It can be thrown and so on. It is generally believed that perception is signal input, but in fact, the computer receives video signal input but has no vision because the computer cannot move. Perception needs to be combined with its own actions, which endows the input signal with semantics. The input signal does not lead to a certain action necessarily. It must be combined with the action. Firstly the production of perception goes through the coordination and integration of input signals, its own motion and environmental objects, and the integration forms an experienced memory. And then when it encounters related signals, it will produce knowledge of objects and actions that can be performed on objects. Of course, only perception is far from enough. Intelligent systems also need the ability to reason, think, and plan. But these abilities can be built on the basis of the perception platform.

The difference in language and information processing between humans and machines is mainly reflected in the ability to connect seemingly unrelated things together. Although the era of big data may change, for machines, the gap between the extraction of abstract representations, that is, the decision-making method based on rule conditions and probability statistics, and the judgment (specific to humans) based on emotional touch and insight meditation, remains to exist.

Einstein once described the difference between logic and imagination like this, "Logic will get you from A to E, imagination will take you everywhere." In fact, the most dominant characteristic of human beings is integration of logic and imagination, concreteness, and abstraction for a purpose according to a specific situation. This kind of flexible diffusion and aggregation mechanism is often related closely to the task situation. Just as when it comes to the concept of words, some philosophers insist that the meaning of words is inherent to the physical objects that exist in the world, and Wirtgen Stein believes that the meaning of a word is determined by the context in which people use the word. The reason is probably due to the phenomenon of competition and risk in a diode-like mechanism. This phenomenon is also present in human's consciousness as if they are indecisive and in a dilemma. The root of ideological struggle is related to uncertainty and the uncertainty of human, object, and situation. Limited rationality may have some connection with it. The key is how to balance and find a satisfactory solution (finding the needle in the bowl) instead of finding the optimal solution (find the needle in the sea). In contrast, in the machine program that recently defeated the world champion of Go, Lee Sedol, the Alphago parameters are adjusted very well. The balance of this parameter is just the critical line of the competitive adventure mechanism, just like the dividing line between the yin and yang in the Taiji picture. There has always been a contradiction between qualitative and quantitative adjustment of parameters in competitive, adventurous behavior. Qualitative is a matter of direction, while quantitative is a matter of accuracy.

For human beings, how did the most mysterious consciousness arise? This question has always been concerned by scholars. There are two main problems: the basic structure of consciousness and the experience accumulated through interaction. The former can be physical or abstract, the difference between human and machines, and the latter is necessary for both humans and machines. Consciousness is the product of human-machine-environmental system interaction. In theory, the current machine has no (active) interaction of the human-machine-environmental system, so there is no reference coordinate system such as you, I, and he. Someone said, "There is no real intelligence in the current artificial intelligence and no knowledge in the current knowledge system. Humans play hard and make it seem logical, natural, convenient, and easy to remember and maintain." This statement is biased certainly. But it also reflects partial truth; that is, consciousness is the product of human-machine-environmental system interaction. The current machine theoretically does not have (active) interaction of the human-machine-environmental system, so it is difficult to reflect a certain order that implies stability and continuity without the reference coordinate system of you, I, and he. The author once communicated with a famous photographer. He once said 10 words to the photographer in a profound way: ① The photo is not well taken because you are not close enough to real life. ② A view captured with the eyes can only be called a photo, and a view captured with the mind can be called art. ③ What I express is my true self, and it comes from the bottom of my heart. ④ Sometimes it is so difficult to obtain the simplest photos. ⑤ There are only good photos but no criteria of good photos. ⑥ The photographer must be part of the photo. ⑦ I think shadow is more attractive to me than the object itself. ⑧ Classic novels, music, and paintings all give me a lot of inspiration. ⑨ I am not fond of using photography as a mirror just reflecting facts, so I leave room for imagination in expression. ⑩ I have been waiting for the interweaving of light and scenery all my life, and then letting magic occur in the camera. These 10 sentences seem to be equally meaningful for the construction of meaning in deep situation awareness.

Sometimes data can be understood or defined as a person's expression or response to a stimulus, just like seeing a word, hearing a sound, etc. Without all kinds of stimuli, intelligence may not be able to develop and grow (not assemble). Einstein said, "Words and language are absent in my mind when thinking. When I think the physical entities are symbols and images and they can be reborn and combined at any time at my own will." Language is the linearization of signs, and language also limits thinking. The difference between human and machine intelligence is the same: the memory type (like machine) and the model (like human). Humans can establish the relationship between the irrelevant factors on a larger scale (even beyond language). The limitation of the machine lies in the limited relevance. For example, describe a system that can track and locate objects in a three-dimensional space. By incorporating the position and direction into the attributes of the target, the system can infer the relationship between these three-dimensional objects. Although, big data may also cause precision interference or cognitive overload (information redundancy is a self-protection strategy in the era of big data). In many applications, small data should also be of great help, because after all, small data is more dependent on analysis. The shortcoming of accuracy is the lack of information redundancy of big data as compensation.

CHAPTER 8

The problem of autonomy

Contents

The future of human-machine hybrid intelligence includes four clear development directions: active recommendation, autonomous learning, natural evolution, and self-immunity. In these four aspects, autonomy is a very important concept.

Integrated Human-Machine Intelligence
ISBN 978-0-323-99562-7
https://doi.org/10.1016/B978-0-323-99562-7.00008-5

1. The development history and status quo of autonomy

1.1 The development history of autonomy

The needs of society are the driving force for the development of automation technology. Automation technology is formed and developed closely around the control of production, military equipment, and the needs of the aviation and aerospace industries.

In 1788, to solve the problem of steam engine speed control proposed in industrial production, Watt connected the centrifugal governor with the steam engine door to form a steam engine speed adjustment system, turning the steam engine into a safe and practical power device. Watt's invention pioneered the research and application of automatic adjustment devices. In the process of solving the subsequent stability of the automatic adjustment device, mathematicians put forward the criterion for judging the stability of the system, and accumulated experience in the design and use of the automatic adjustment device.

The 1940s was a critical period for the formation of automation technology and theory. To solve the technical problems of artillery control, torpedo navigation, and aircraft navigation proposed in the military field, a group of scientists gradually formed a classical control theory and method that regards the analysis and design of the single-variable control system as the main content. The development of mechanical, electrical, and electronic technology provides technical means for production automation. In 1946, Harder, a mechanical engineer at Ford in the United States, first proposed to use the term "automation" to describe the automatic operation of the production process. In 1947, the first production automation research department was established. In 1952, Diebold published his first book *"Automation"* He believed that "automation is a means of analyzing, organizing and controlling the production process" In fact, automation is the result of applying automatic control to the production process. After the 1950s, automatic control began to be popularized and applied as an important means to improve productivity. Its application in mechanical manufacturing formed automation of mechanical manufacturing. It was applied in continuous production processes such as petroleum, chemical, metallurgy, and so on. The production equipment is controlled and managed to form process automation. The promotion and application of electronic computers combine automatic control with information processing, and business management automation appears.

From the 1950s to the early 1960s, a large number of engineering practices, especially the development of aerospace technology, involved a large number of optimal control problems for multiinput and multioutput systems, which were difficult to solve with classical control theories. Thus modern control theory appears with the core of maximum value principle, dynamic programming, and state space method. Modern control theory provides a control method that satisfies the launch of the first artificial satellite and ensures the implementation of several subsequent space programs (such as missile guidance and spacecraft control). The transfer used to only consider the input and output relationship of the control system based on the transfer function, but now it focuses on the state space method to consider the internal structure of the system, which is a leap in the control worker's understanding of the laws of the control system.

After the mid-1960s, the application of modern control theory in automation, especially in the fields of aviation and aerospace, produced some new control methods and structures, such as adaptive and stochastic control, system identification, differential strategy, and distribution parameter system, etc. At the same time, pattern recognition and artificial intelligence have also developed, and intelligent robots and expert systems have emerged. The application of modern control theory and electronic computers in industrial production has enabled the control and management of the production process to be integrated and optimized.

In the mid-1970s, automation began to be applied in large-scale and complex systems, such as large-scale power systems, transportation systems, steel complexes, and national economic systems. It not only requires optimal control and management of existing systems, but the optimal planning and design of the future system is also required. The use of modern control theory and methods can no longer achieve the desired results, so large-scale system theories and methods have emerged. With the rapid development of computer networks in the early 1980s, greater progress has been made in management automation, and management information systems, office automation systems, and decision support systems have emerged.

With the development of the times, human beings have begun to solve the problems they face by comprehensively using new technologies and new methods such as sensing technology, communication technology, computers, system control, and artificial intelligence. Automation can no longer meet people's needs. To further reduce the demand for human resources, reduce the burden on people, reduce dependence on high-bandwidth data links, shorten the cycle of task loops, and improve the ability to complete tasks autonomously, the demand for autonomy in unmanned systems is becoming stronger. Autonomy is not only a typical feature and development trend of unmanned systems, but also an advanced stage of automation development.

1.2 The status quo of the development of autonomy

1.2.1 Sense and percept

The ability of sense and percept is the key element to realize autonomy. Only through sense and percept can unmanned platforms reach the target area and achieve mission goals. For example, the platform collects sensor data, applies kinetic energy weapons, and fights improvised explosive devices, which are all inseparable from the capabilities of sense and percept.

According to the purpose of the ability of sense and percept, people divide the sense and percept of unmanned systems into four categories, namely navigation sense and percept, task sense and percept, system health sense and percept, and operation sense and percept. To accomplish a certain task, these four categories often overlap.

When starting the guidance, navigation, and control functions, it is necessary to support path planning and dynamic replanning through navigation's ability of sense and percept to achieve multiagent communication and coordination. Navigation generally refers to the whole process of the platform toward the target direction, which is the opposite of platform motion control (such as maintaining a vertical position or choosing footwork for a footed robot). By improving the navigation's ability of sense and percept, the safety of the platform can be improved (because human response speed is usually not fast enough, and the lag of the network cannot be overcome, so the reliability and safety of navigation cannot be guaranteed), and the cognitive workload when operating the platform or driving the platform is reduced, although this is not enough to reduce manpower requirements. By selecting the airborne sense and percept processing method, the response speed of receipts can be improved, and the platform can help the platform fight against network attacks or network destruction.

Task planning, scenario planning, evaluation and understanding, multiagent communication and coordination, and situation awareness all require the support of task percept. Improving the autonomous ability of sense and percept of task sense and percept can bring the following four major benefits.

(1) The robot can perform tasks in secret, for example, tracking an activity without the need for a full network connection, thereby reducing the possibility of network attacks and reducing the cognitive function load of the operator.
(2) Through active identification (even if the target is prompted or the given target is prioritized), the demand for data analysts can be reduced.
(3) By airborne confirmation or prioritizing some of the data to be sent, network requirements can be reduced. For example, "Global Hawk" requires a lot of bandwidth.
(4) Mission perception can be combined with navigation: for example, only the platform hovers, keeps stationary, or spins in the air.

Platform health sense and percept is mainly used in fault detection and platform health management, but it is also in performing fault sense and percept, replanning, and accident management. Strengthening independent health monitoring has at least the following three benefits.

(1) When the speed of autonomous fault detection, confirmation, and repair may be higher than the speed of manual detection, confirmation, and repair, the fault would be weakened and even repaired.
(2) It can increase the user's trust in the system, especially if the system does not operate as expected or suddenly fails during the critical phase of the mission.
(3) It can further reduce the cognitive workload of the operator, and it is no longer necessary to specially arrange an operator to monitor the diagnosis display throughout the process.

As the navigation location shifts from outdoors to indoors, and the mission focus shifts from remote sense and percept to travel actions, operational perception becomes more and more important. For example, using a ground robot to open a door is an arduous task. In addition, other tasks that need to be done with operational sense and percept include dismantling improvised explosive devices, vehicle inspections (in the process, packages and other objects need to be moved), and logistics and material handling, etc. Improving the sense and percept of autonomous operation has the following two major benefits.

(1) The time and workload required to complete the operation task can be reduced.
(2) The number of robots participating in the task can be reduced, because it is usually necessary to arrange a second robot to assist the operator in

monitoring the relationship between the manipulator and the operated object at any time before improving the ability of autonomous control sense and percept.

1.2.2 Planning

Planning refers to the calculation process of the sequence of actions or partial order that can change the current state to the expected state. Under the premise of using as few resources as possible, the process is acting to achieve the mission objectives. In this process, there are two key points.

(1) describe actions and environmental conditions, set goals or resource optimization standards;

(2) provide algorithms for computing action sequences and allocating action resources under the premise of complying with hard conditions (for example, the platform's conditional restrictions on terrain and speed, etc.) and optimizing soft constraints (for example, minimizing the time or manpower required to complete the task).

1.2.3 Learning

Machine learning has now become one of the most effective ways to develop intelligent autonomous systems. Generally speaking, it is more efficient to obtain information autonomously from data than from manual knowledge engineering. The latest technology systems of computer vision, robotics, natural language understanding, and planning mainly rely on training data for autonomous learning. Finding a reliable model from a large amount of specific data can generally make the accuracy and robustness of an autonomous system higher than that of manual software engineering, and it can also make the system automatically adapt to the new environment based on actual operating experience.

1.2.4 Human-machine hybrid

The human-machine hybrid is a relatively new interdisciplinary field. It mainly solves the problem of how humans and robots, computers, or tools collaborate. It is a branch of the field of human-system interaction, focusing on the two-way cognitive hybrid relationship between the humans and robots. In this human-machine hybrid relationship, the robot assumes the role of the agent and runs in a location far away from the user, calculation, or autopilot, which has obvious technical advantages.

Human-machine hybrid covers many fields such as unmanned systems, human factors, psychology, cognitive science, communications, human-machine hybrid interaction, computer support working groups, and sociology. This huge interdisciplinary state is obviously different from those of the traditional engineering design, interface development, or bioengineering.

Studying the relationship between the man-machine system and the platform will help improve system performance, reduce platform operating costs and design costs, improve the existing system's adaptive ability to the new environment, and speed up its progress. By improving the collaboration relationship between humans and unmanned platforms, the system can perform tasks faster and reduce the error rate at the same time; also, by improving the collaboration between humans and the platform, the communication interface is improved, and the availability and reliability of the application are improved. It can reduce the demand for system operators and reduce the cost of designing different system displays or redesigning unmanned systems in the absence of human-machine hybrid interaction support. A better understanding of the roles and limitations of humans, unmanned platforms, and autonomy under special circumstances will help design a system that can not only monitor transcendence but also predict new demands, thereby improving the adaptive ability of system performance. By improving the level of human-machine hybrid, the task execution ability of the unmanned system would be improved, and the trust of humans in the system can be improved. In addition, using the advanced human-machine hybrid anthropologic methods, we can capture innovation opportunities during the use of unmanned systems, thereby accelerating capabilities, new uses, and optimal implementation.

As shown in Fig. 8.1, human supervision on the loop includes the following.

(1) allowing to act without the constraints of the design (behavior 2);
(2) allowing to operate outside the combat situation (behavior 3);
(3) effective use of opportunities for dynamic changes.

1.2.5 Natural language understanding

Natural language processing is closely related to the development of computing systems that can communicate with humans in common languages such as English. Automatic speech recognition is the process of converting speech signals into text information, while natural language understanding is the process of converting text information into formal expressions that computers can understand. The human-machine hybrid is affected by natural language. If all the instructions sent to the unmanned

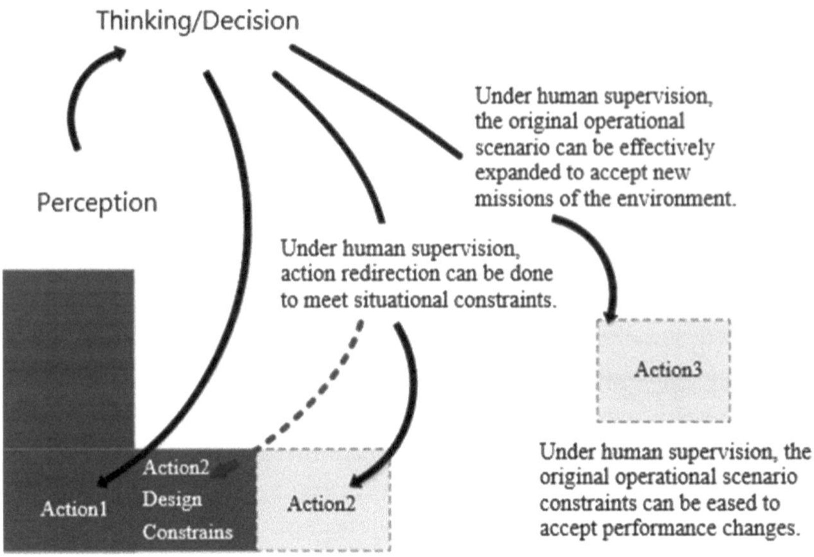

Figure 8.1 Perception—behavior—decision diagram.

system are accurate, the content of the delegation will be simplified, and the speed of the delegation will be increased accordingly. However, natural language is an independent research subject.

When a person issues instructions to an autonomous system, natural language is the most normal and natural way. Humans generally use natural language processing to customize diversified high-level goals and strategies for self-service systems, rather than directly performing specific remote-control operations. However, due to the inherent uncertainty of natural language processing, there are some difficulties in the understanding of natural language processing. When understanding natural language, it is necessary to judge the true meaning of the language in combination with the context at that time. Building an autonomous system that can understand English instructions and human language will be a highly technical challenge. To solve this problem, we often use traditional graphical user interfaces to communicate with computer systems. However, in most cases (for example, when it is inconvenient for the user to operate by hand), language is still the most ideal way to communicate.

1.2.6 Multiagent coordination

When performing tasks across robots or software agents and natural persons, people often refer to the term "multi-agent coordination" Each agent has a certain degree of autonomy. There are two ways to coordinate between multiple agents, namely distributed coordination and centralized coordination. Distributed coordination refers to the direct interaction or negotiation of multiple agents, while centralized coordination refers to unified coordination under the guidance of the planner. No matter which method the agent uses to coordinate, people must ensure that the agent can not only synchronize but also adapt to dynamic changes in the environment or tasks. Multiagent synchronization is often understood as active collaboration (such as robot soccer) or nonactive collaboration (such as ant foraging behavior) between multiple intelligent systems. Although there is a certain connection between collaboration (human-machine collaboration) and collaboration, it refers to completely different topics. It assumes that each agent has a certain cognitive understanding of the capabilities of other agents and can accomplish the goal. The progress is monitored, and the formation can be carried out like a human. Therefore, in the research process, multiagent coordination and human-machine hybrid interaction are two interrelated technical fields, but multiagent coordination research mainly focuses on agent coordination mechanisms with different configurations, while human-machine hybrid interaction focuses on collaborative cognition.

Multiunmanned platform coordination has at least four major benefits: expanding coverage, reducing costs, providing redundancy, and achieving standardization. Compared with the independent work of a single platform, the coordination of multiple unmanned platforms has a wider common coverage and stronger endurance. It can not only play the role of network communication relay but also provide guarantee for the coverage of sensor networks. Multiple low-cost unmanned platforms can replace a single high-cost and low-cost observable platform, and they can also replace the high-level protection systems necessary to deal with "anti-access" and "area denial" In the presence of noise, chaos, interference, camouflage, concealment, and deception, multiple low-cost platforms can provide redundancy in parallel (even if several of them are performing other tasks or failing), and they can still complete the task in the end. By coordinating multiple dedicated platforms or heterogeneous platforms, costs can be reduced, and design requirements can be reduced too.

Through autonomous coordination, multiple unmanned platforms can quickly complete coordination and optimization, reduce error rates, and reduce or eliminate dependence on network communications or other resources. Using autonomous planning capabilities, unmanned platforms can be optimized under conditions of dynamic changes. Planning and scheduling algorithms can coordinate thousands of agents and constraints in real time, which humans cannot achieve by themselves. Coordination is not limited to the movement planning of parallel activities, but it also includes coordinating a series of activities. For example, to observe from multiple spectrums or multiple perspectives (such as air to ground), tasks are assigned to dedicated unmanned platforms through general unmanned platforms. The realization of autonomous coordination does not necessarily require network communication. Therefore, unmanned platforms can also be used in hidden areas, simulated environments, or regions without communications.

2. The concept of autonomy and its theoretical origin

2.1 The concept of autonomy

Autonomy is derived from ancient Greek, meaning "the person who gives law to himself" which is a concept in moral, political, and bioethical philosophy. A rational individual has the ability to make an informed, noncompulsive decision.

In the field of sociology, the debate on the boundaries of autonomy has remained on the concept of relative autonomy until the creation and development of autonomy classifications in scientific and technologic research. It is believed that the autonomous form of contemporary science is reflexive autonomy: actors and structures in the scientific field can translate or reflect different themes put forward in the social and political fields and influence the topic selection of research projects.

The philosopher Ian King proposed a "principle of autonomy" on how to make the right decisions and always maintain the right attitude. He defined it as, "Let people choose for themselves unless we understand their interests better than them."

The Swiss philosopher Jean Piaget (1896–1980) analyzed children's cognitive development during play and confirmed through interviews (among other principles) that the process of children's moral maturity is divided into two stages: the heteronomy stage and the autonomy stage.

Heterologous reasoning: The rules, which are objective and unchanging, must be literal because they are ordered by authority and are not suitable for exceptions or discussions. The basis of this rule is the authority of superiors (parents, adults, the country), and no reason should be given for implementing or fulfilling the rule under any circumstances.

Autonomous reasoning: The rules are the product of the agreement, which can be modified. They can be explained and are suitable for exceptions and objections. The basis of this rule is its own acceptance, and its meaning must be explained. Sanctions must be commensurate with the absence, assuming that criminal acts can go unpunished sometimes, so it is unacceptable if collective punishment is not guilty, and the offender cannot be punished under this situation.

In the field of medicine, respect for the autonomy of patients is regarded as one of many basic ethical principles in medicine. Autonomy can be defined as a person's ability to make their own decisions. This belief in autonomy is the core premise of the concept of informed consent and joint decision-making. The seven elements of informed consent include threshold elements (capacity and voluntary), information elements (disclosure, advice, and understanding), and consent elements (decision and authorization).

There are many different definitions of autonomy, many of which place individuals in a social environment, such as relational autonomy, which means that a person is defined through relationships with others, and "support autonomy," which means that it may be necessary to temporarily impair the person's autonomy in the short term under certain circumstances to maintain their autonomy in the long term.

In the field of robotics, autonomy or autonomous behavior is a controversial term referring to unmanned systems (such as unmanned cars), because people cannot understand whether decisions are based on their own decision-making abilities or on the preprogrammed decisions. This is an abstract quality that is difficult to measure. In a sense, the autonomy of the machine is just an analogy, which does not include the ethics of human society, and automatic means that the system will operate exactly as programmed, and it has no choice. Autonomy means that a system can choose not to be influenced by the outside world; that is, an autonomous system has free will. A truly autonomous system is capable of accomplishing complex tasks without internal human guidance. Such a system arguably further automates other parts of the entire process, making the entire "system" larger, including more devices that can communicate with each other, without involving people and their communications.

In mathematical analysis, it is called an autonomous equation if an ordinary differential equation is independent of time. In linguistics, an autonomous language is a language that is independent of other languages, such as standards, grammars, dictionaries, or literature, etc. In robotics, autonomy means independence of control. This property means that autonomy is a property of the relationship between two intelligent agents, and in robotics, it is a property of the relationship between the designer and the autonomous robot. According to Rolf Pfeifer, self-sufficiency, positionality, learning or development, and evolution increase the degree of autonomy of an intelligent human. In space missions, autonomy can also refer to manned flight performed without the control of a ground controller task. In social psychology, autonomy is a personality trait characterized by an emphasis on independence for personal achievement and a preference for solitude, often labeled the opposite of social orientation.

Autonomy can be defined as a kind of action that acts without waiting for external force, which can form a beneficial situation and make things go as expected. Some people define autonomy more simply as being opinionated and acting in their own way.

In a traditional way, automation is defined as a process of equipment and systems completing specific operations and realizing the expected goal without or with a little manual participation. In a general way, automation includes applications and other applying processes that are meant to execute logical steps and practical operation.

Autonomic systems refer to those systems that have self-led and self-managed abilities and can deal with nonprogrammed and nonpresent situations. Compared with automated equipment and autonomic systems, autonomic equipment and autonomic systems can reply to the more complicated environments, complete more operations and control, and have broader potential applications. Generally speaking, autonomy means the process that uses application sensors and complicated software to complete tasks independently without outer interference, keep the device from communicating and doing self-regulation in an excellent continuous condition.

Nowadays, most unmanned aerial vehicles (UAVs), unmanned ground vehicles (UGVs), and autonomous underwater vehicles (AUVs) need manual control in a low level of autonomy. In the future, those remote control equipment may well include more automatic functions such as remote control, semiautonomy, and full autonomy (actually semiautonomy in some aspects). In the future, autonomy is the eventual destination of the control field. But for a long time, with the development of autonomic systems, most tasks, including commanding control and coordinating action, need to be accomplished with the help of people. Man-machine hybrid intelligence is the unity of relativity and absoluteness.

2.2 The origin of autonomy

The autonomy comes from that the admitted entity is delegated to policy, taking action within the given limit. An obvious difference between autonomy and automation is that the system managed by rules, which are not admitting any deviation, is an automated system instead of an autonomous system, and an autonomous system must be capable of making and choosing different action processes to understand and realize related targets according to knowledge about the real world, system, and situation.

Autonomous systems mainly come from artificial intelligence, which is the ability of computer systems to perform tasks with the participation of human wisdom. Thanks to the continued development of AI, the human can delegate tasks to the machines once machines cannot perform them.

The artificial systems are aimed to apply AI in specific areas or solving special difficulties. Specially speaking, systems work within the boundaries of the defined knowledge base after programming trains. Autonomous functions are at the system level instead of the structure level. We mainly consider two structure systems, application static systems and dynamic application systems. In a broad sense, systems applying static autonomy are actually working through software, including planning and professional consulting systems, but systems applying dynamic autonomy will enter the material world, robots and autonomic platforms included.

The robot technology promotes the development of new sensors and brakes, and it advances the movement of artificial systems at the same time. Generally, early robots are automatic robots. In recent years, autonomous functions have gradually been improved a lot with the development of AI.

3. The theoretical expression and model of autonomy

3.1 The meaning of autonomy

"Autonomy" is driven by information, even knowledge. Unmanned systems autonomously complete the dynamic process of "feeling—judging—deciding—acting" according to tasks and needs, react to unexpected situations and tasks, and tolerate failure to some extent.

Autonomy gradually changes the world through several applications such as data searching and analysis, internet searching, advice engines, expecting, and so on. Considering the limited ability of humans to handle to huge data quickly, autonomous systems can be utilized to discover the trends and analyze models.

In more complicated situations or environments with more multiple tasks and actions, more sensors and complicated software are used, providing autonomy at a higher level. The features of autonomy are usually illustrated by the level of systems completing tasks autonomously. In other words, autonomous systems should be able to eliminate outside disruptions in extremely unsure conditions and still compensate for problems that system failure brings in the case of bad communication and make sure that the system can perform well for a long time.

3.2 The build of an autonomous system framework

In the process of designing the autonomous system, much energy has been spent to decide whether computers or operators play specific cognition functions. These decisions reflect the balance of different capability elements at the systematical level. For example, while facing the problems of expectations, the effective and best solutions are found at the calculation level, but solutions may fail when expectations change or a new situation occurs. What is more, it is extremely sensitive to add more human sources. On many occasions, obeying those absolute design decisions does not need to examine the influence on system end users or overall communication and maintenance or labor cost.

In the process of some project practices, it has been discovered that dividing autonomous levels is not obviously helpful to the autonomous design. These projects have cost too much energy at the computer level and have not focused on a collaborative relationship between operators and supervisors to maximize capabilities and effects, so these projects do not perform well.

Either in cognition science level or according to observation results of actual practice, these classification systems are misleading. At the cognition level, when many functions are completed by computers, only advanced monitoring or supervising are completed by operators. Actually, all decisions, which are continuously unified, are under the control of humans. In some conditions, many functions may need to be performed at the same time to indicate that systems have some specific capability, some of which need humans yet others will be delegated to computers. So in any stage of a task, the system has the possibility to stay at two or more unrelated level. In a realistic process, because some people regard "autonomous grade" as routine development, only machines instead of the human-machine system need to be paid attention to.

Autonomous system models mainly have these points: focusing on human-machine function and redistribution of decisions to realize specific capability, where the way of distribution differs from each other in different stages and at different cognition level of tasks, and advanced system balance must be performed while designing visual autonomous capability. The autonomous system designs and estimation framework are shown in Fig. 8.2.

(1) Views at a cognition level

As the autonomous level of component agents develops and functions enhance continuously, it is more important to perform unified action to coordinate at different levels functions.

Cognition-level view mainly considers that autonomous technology is used to regulate a "user" control range and extend it to other space to promote adaptability. Platform action, sensor operation, communication, and state monitoring are controlled by platform or sensor operator, and departments or formation leaders take charge of task planning, replanning, and cooperation among multiple agents. The control range of task commander or executive officer includes imagination assessment and understanding, imagination planning and deciding, and management with unexpected incidents. What is more, communication and cooperation are needed among operators. Every cognition function can be distributed between calculation and operator or supervisor or undertaken by both calculation and supervisor.

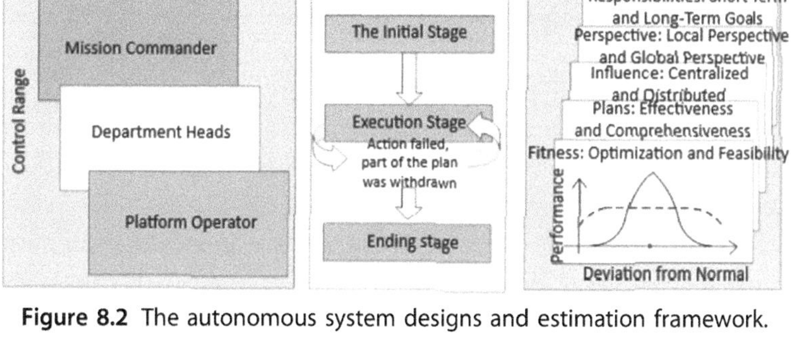

Figure 8.2 The autonomous system designs and estimation framework.

(2) Task dynamic view

In the different stages of a tasks, autonomic technology is applied in different ways. This view also reflects ways of different agents in different stages, with different functions and at different levels achieving action synchronization when a new accident, controversy, or change occurs.

According to the task dynamic view, the distribution of cognition function will change when tasks based on the environmental complexity and necessary response time are performed. Promoting autonomy level favors the adjustment of plan according to requirements in the process of tasks, for example, a new purpose appearing or changing, additional information, deterioration of conditions, or decrease of platform performance level. In the initial and final stage, it becomes possible to utilize autonomous technology to decrease manpower and increase efficiency.

(3) View of complicated system balanced space

Design choices about places and ways of autonomous technologic application can change the way of a huge system to weigh and balance multiple performances. There exist risks because it is possible that other areas in the comprehensive performance are affected badly if autonomous technology only improves one area.

A view of complicated system balance space should make the right trade-offs according to the five aspects listed next.

① Adjustment

Trade-offs should be made between self-adjustment ability and performance optimization of new tasks and unexpected situations.

② Plan

Trade-offs should be made between needs that must be changed for ineffectiveness when following some current plan process.

③ Influence

Trade-offs should be made between centralization and distribution to enable information, either accessed remotely or locally, to be visible under the premise of not being influenced by potential or unknown elements.

④ Vision

Trade-offs should be made between locality and wholeness to master the situation and help to adapt the concentrated action in one unit to the distraction and cooperation in many units to realize better effect.

⑤ Responsibility

Trade-offs should be made between long-term and short-term purposes and reach an agreement. A complicated system balance space view is shown in Fig. 8.3.

Space of Tradeoff	Object of Tradeoff	Effect	Adverse Consequences
Adaptability	Optimality and feasibility	You can get better results when you see the situation clearly	Flaws increase
Planning	Effectiveness and comprehensiveness	Achieving balanced use of computing resources	Causing planning errors or difficulty in revising plans
Influence	Centralization and distribution	Adapting clipping actions to appropriate layers	Coordination costs rise
Perspective	Local perspective and global perspective	Adapting the scale, scope and resolution of the action	Causing data overload, slow decision-making
Responsibility	Short-term and long-term goals	Building trust and aligning branch management with mission goals, priorities, and context	Cause failures of collaboration or coordination

Figure 8.3 Complicated system balance space view.

4. The consideration of autonomous problems

4.1 Human's decision and machine's decision

We are giving more and more problems to computers, but we do not know whether a machine's decision is better than a human's. The original intention of making decision through a computer must be kind, namely increasing efficiency, winning data support for decisions rapidly, and making process clear. However, when wondering at those amazing autonomous decision system algorithms, people always ignore an important problem that autonomous decision systems may bring much negative influence because of data deviation, a system built-in defect.

Autonomy is usually used to support people's decisions. Expert systems or decision support systems provide decision knowing, for example, post-elevation of action assess, target prompt, classification on detected target, and so on. Actually there is difficulty with effective support. Although we usually assume this kind of system could improve a human's decisions, especially when carrying out a difficult task, the fact is not like that. It is proved that humans often accept system assessment and advice first, and then combine it with knowledge and understanding on a situation. So wrong assistance will bring wrong decision, which will increase the chance of error.

What is more, because the information source increases, the deciding time increases too. As a result, chances are that the accuracy and real time of a human or machine decision system will not appear if there is defect in the assisting decision system. Good advice is helpful, but if advice deviates from accuracy, the decision maker will make mistakes, which causes the decrease of the whole task performance.

On the contrary, the decision support system targeted at assessing a human's decision can propose a deviation of scheme aiming at solving difficulty from human to computer because the input happens after the human's decision. It also utilizes an advantage of a computer that imitates a human's solution to the situation quickly, which is good for distinguishing potential disadvantages, promoting cooperation between human and machine, and increasing whole performance.

4.2 Individual situation awareness and group sharing situation awareness

As the autonomous ability grows stronger, the intelligent level is improved, and it becomes able to deal with more situations and functions. Then operators are needed to improve the ability to understand current work to ensure the interaction with the system. As for the future autonomous system, it is necessary to develop an advanced interface to support needs to share situation awareness between operators and autonomy. Sharing situation awareness is the key to support multiparty collaborative action (having the same target and relative functions).

Sharing situation awareness means that "the operators are at the same level of situation awareness and have shared situation awareness based on needs" namely, the shared situation information needed by both sides. If operators are all human, there are still challenges on the issues of getting shared situation awareness, even though getting the same input from the monitor and in the same environment. Because their purposes, formed systems, and environmental mental models are all different, the information will be decoded in different ways, and the future protection will be different.

Autonomous systems utilize computer models to explain the information from sensors and input, so autonomy and its crew may well have different opinions over the real environment that affect their decisions. To deal with this challenge, the crew and autonomy should provide effective situation models to communicate mutually. This means not only both fundamental data but also the way of data decoding and each future prediction should be shared.

Group autonomy is required to realize advanced sharing situation awareness (Fig. 8.4) to support the basic multiple operation related to this concept, as follows.

① Target

Crew and autonomy should support the same dynamic changing target. For example, the autonomy is to land at the airport; if the target of a pilot is to re-fly, then the problem appears. Prison level and targets will change, and sharing situation awareness should make sure the target of both autonomy and crew is cooperative.

② Function distribution and redistribution

Flexible autonomy needs to continuously distribute functions to crew and autonomic system. The main part and object of action, and related abilities and situations for crew and autonomy should be known to carry out different functions.

③ Decision communication

When crew and autonomy are deciding on how to carry out different functions, it is necessary that these decisions (including strategy, plans, and

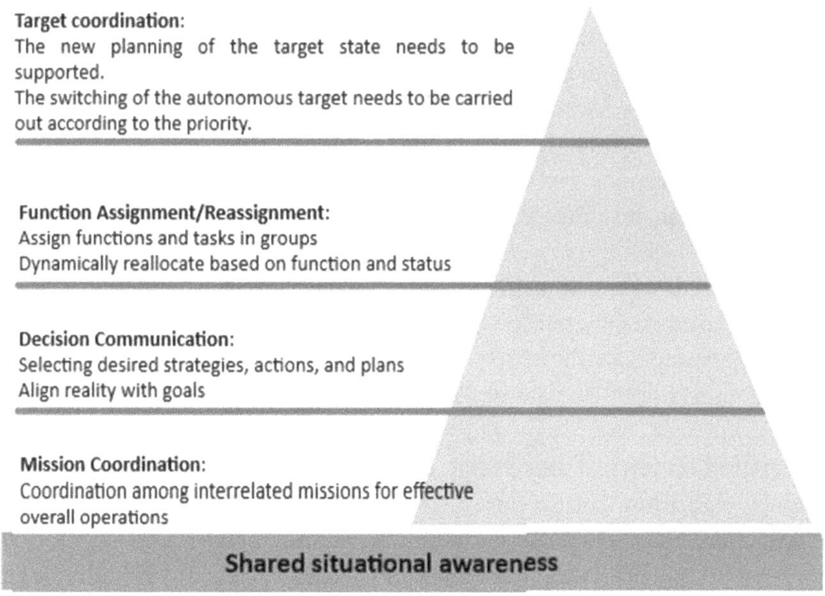

Figure 8.4 The architecture of shared situational awareness.

actions) are paid attention to. What is more, mutual sharing and designing related functions are realized.

④ Task coordination

Autonomy and crew may have highly relevant tasks, which often rely on each other. Both sides need to keep understanding over action of the other side and the level of sharing.

In the process of handling a large quantity of information about sharing situation awareness, gathering information under the circumstances of obeying a principle may cause phenomenon such as "$a > b$, $b > c$, $c > a$." Obviously, the results, according to lots of information, might go against the logic that "it is impossible that $c > a$" in the individual situation awareness. So it is necessary to utilize group sharing situation awareness legally and combine it with individual situation awareness to make use of sharing information to the most extent and decrease absolute dependence.

4.3 How to build human-machine trust

Trust is complicated and multidimensional. When a human assigns systems deploying decisions for a given task, they must trust those systems, so do all the stakeholders who have influence on many other decisions. It is a basic requirement of operators, commanders, designers, testers, policy or law makers, and even the public that one builds credibility and provides proper instruction ability to ensure assessment of unavoidable changes when designing.

Proper design and execution of a system, high ability, high reliability, high wholeness, and so on are the methods to ensure trust. Of course, designers should embed those attributes when designing and manufacturing autonomous weapons. However, these are influences that may be affected by multiple human-machine features in formation easily. They are as follows.

1. Machine lacks humanoid percept and thinking

Compared with a human, autonomous systems have different sensors and data sources. Therefore, they may be used under the condition of different assumptions about combat circumstance. What is more, machines may adopt methods that are obviously different from those of the deciders to reason with regarding specific algorithm options, such as pattern recognition in image processing, decision optimization algorithm, deep learning algorithm, and so on.

2. Machine lacks self-percept and environmental percept

Percept may be as simple as an understanding of a healthy situation itself, such as battery level, or as complicated as perceiving when to be used beyond original design bounds or assuming conditions. Environment percept includes traditional environment percept, such as communicating under the circumstances of wing icing or interference, and other complicated effect such as lying GPS. Of course, to machines, it is far from enough to perceive the change of itself and the environment. It is also required to adapt to these changes flexibly and effectively on this basis.

3. Observable attribute, predictable attribute, indicative attribute, and interrogating attribute

An autonomous system is not only required to be in motion within its power under the circumstances of dynamic and complicated war, but also to send related and visible information to humans and other machines. What is more, even if machines can make sure the current situation and effect are visible, there is still possibility that expected indicators are not enough to ensure predictability. Besides, once mistakes appear, autonomous systems must have the ability to ensure that humans and other machines intervene, correcting or ending the errors to keep indicative. At last, machines should have an interrogating attribute. In other words, machines must save and provide an understandable and unchangeable record about inference behind decision and action according to facts.

4. Understanding on the sharing target of human and system is not enough

To ensure cooperation effective between any autonomous systems, common targets should be set and understood.

5. Invalid interface

The speed of traditional computer interfaces that achieve communicating between human and machine, such as mouse clicks, is slow to communicate between human and machine, which hinders cooperation in the time-sensitive and dangerous condition. Improving the interface helps relieve these problems.

6. Systems that have the ability to learn

Currently developed machines can generally change their own capabilities and constraints and adapt to their uses and environment. This system will go beyond its initial verification and validation and require a more dynamic method to ensure the effective implementation of relevant tasks in the whole life cycle. The operator must be able to perform the task independently for half a period of time. The degree of approximation is related to the reliability of the whole system and the performance evaluation of the system under special situation. Therefore, the crew must establish informed trust, accurately evaluating the timing, degree, and intervention time of the application of autonomy, and calibrating the degree of trust, such as too much trust, trust, lack of trust, etc.

5. The future developing and investment direction

5.1 Percept

Development direction is the environmental percept and situation understanding in a complex battlefield. The percept plan mainly focuses on improving the navigation autonomy of a single platform or related platform groups, while platform task perception is secondary.

The specific points of development are as follows.

(1) The comprehensive application of a single-platform battlefield perception has no obvious advantages, compared with map annotation, and the role of percept in helping combatants master the operation status of the platform and presenting the relationship between the platform, battlefield, and mission objectives is often ignored. On the contrary, it is also wrongly regarded as a problem of computing and display. However, display cannot make up for the lack of perceptual ability.

(2) Eliminate the airspace conflict of manned and unmanned system intensive operation. Perceptual avoidance technology has passed the inspection, and there are many schemes available. However, the main development direction lies in the lack of basic theory, rather than consolidating these schemes, integrating them into the existing technical schemes, and meeting the constraints of social organizations.

(3) Perform real-time detection and confirmation of sudden threat sources. Threat source detection and confirmation is the highest level of situation understanding. In this mode, the fighter can identify and design necessary actions. When detecting threat sources, we can not only use the airborne sensing system with a single platform, but also integrate the observation intelligence of multiple platforms and information from other channels to ensure the success of the detection mission.

(4) Perform high-speed detection of obstacles in complex terrain. The research on the technology of navigation for unmanned ground platforms in cities, wild forests, or people is still in its infancy.

(5) For multisensor synthesis, there is usually a one-to-one relationship between the sensing capabilities of unmanned systems and sensors.

Although multisensor synthesis can improve the reliability of sensing and the comprehensiveness of environmental modeling, it is often ignored. In addition, we need to weigh the investment in percept and sensor, reliable percept and evidence reasoning ability of platform health monitoring, as well as operation perception. Do not pay too much attention to the development of new sensors and ignore the optimization of existing sensor algorithms. Do not divide percept into human percept and computer percept, nor think that there is no cross-relationship between these two categories and ignore the cooperative relationship of man—machine collaborative perception. Do not put self-percept above manual performance.

5.2 Planning

Development direction is making up for the deficiency of an autonomous system and the user's own knowledge to the greatest extent.

Specific points of development are implementation monitoring and replanning. There is a famous saying, "No battle plan is effective after meeting the enemy." Therefore, the key to the success of the action is monitoring the overall situation, detecting faults, capturing opportunities, and adjusting the plan to adapt to the form while carrying out the planning work.

5.3 Learning

Development direction is adapting to an unstructured dynamic environment and developing learning technologies that can adapt to these complex environments.

Machine learning has become one of the most effective ways to develop intelligent autonomous systems. In general, getting information from data is more efficient than manual knowledge engineering.

Specific key points of development are adopting a variety of technologies to reduce the supervision workload of a learning system, including active learning, transfer learning, semisupervised learning, cross-modal training, reinforcement learning, imitation learning, and so on.

5.4 Interaction between human and machine

Development direction is good communication between people and the unmanned platform; work modeling between human and unmanned platform; researching and improving the cooperation between human and unmanned platform; predicting the availability and reliability of human unmanned platform collaboration; and capturing and expressing the interactive relationship between human and unmanned platform in special field applications.

Portraying end users.

Key points of development are achievement of the natural user interface required for trusted person system cooperation, as follows:

(1) Operator controls interface can quickly complete the system, conventional sensors, and viewpoint-related training under the condition of multiple degrees of freedom, to cultivate novices into professionals.

(2) Perception-oriented interfaces and sensors are designed according to the psychological and physiologic characteristics of a human body notification system.

(3) The interface shows the current operation of the unmanned system and the correlation between the expected mission objectives.

(4) Human-system effective dialogue uses human-machine natural interaction mode, especially natural language and font action.

The understandable behavior of an autonomous system is as follows.

(1) For the human-machine interaction protocol model of an unmanned system, using the specification model, you can specify design criteria, evaluation criteria, and operation test and evaluation procedures.

(2) System knowledge or system state model specifically designed for operators or decision makers can maintain the credibility of the predictable results of the system.

(3) Cost benefit data collection/analysis methods will help to improve the understanding of the field operation mode of unmanned system and its autonomous ability.

5.5 Natural language understanding

Developing direction is that operators only need to send oral instruction to complete the task according to the information that is received by visual attention mechanism, thus decreasing working load and improving survival rate of personnel under bad circumstances.

Nowadays natural language understanding focuses on understanding literal text but ignores instruction and dialogue understanding that focuses on actual circumstance interaction.

Developing points of natural language understanding are as follows.

(1) In explanation of situational language, words and sequences are connected with percept of objects and incidents in real life.
(2) For understanding over instructive language, the natural language instructions are mapped to normal action sequences performed by robots.
(3) Understanding of space language explains the language expression of space relation in the environment.
(4) Situational dialogue refers to the mixed and active natural language dialogue, which means to achieve human–machine interaction and cooperation.

5.6 Multiagent coordination

Development direction is that each agent has a certain degree of autonomy. Multiagents can coordinate in two ways, namely distribution coordination and centralization coordination. Distribution coordination refers to that agents interact or negotiate directly. Centralization coordination refers to unified coordination under the guidance of the planning period. No matter how the agent coordinates, it must be ensured that the agent not only synchronizes but also adapts to dynamic changes in the environment or task.

Specific development points are as follows.

(1) There is normalized mapping of appropriate coordination schemes and system attributes for specific types of tasks. Up to now, the research work of a multiagent system has still adopted an ad hoc network, mainly focusing on the development of new coordination algorithms, ignoring the development of new application fields, and it does not pay attention to integrating the research results into the specification of the exclusive design theory, so designers cannot select the most appropriate system for a specific task. Although the current classification system can only play a fundamental role in this issue, we must strengthen our work in this area in the future.
(2) In demonstrably correct emergent behavior, both weakly coordinated conscious systems and weakly coordinated unconscious systems use biologic algorithms to minimize the amount of communication, calculation, and perception. There are many advantages to a fleet of low-cost unmanned platforms. But there are no tools in place to predict

the consequences of a sudden change in the environment or whether unmanned platforms will behave correctly in response.

(3) For interference and opportunistic task reassignment, if multiple robots cooperate under the guidance of the common goal, there may be unintentional interference between the robots, resulting in a decline in work efficiency. More significantly, capabilities can be shared and distributed in real time if multiple collaborative unmanned platforms are part of one system locally and part of another globally, often collaborating spatially.

(4) Communication includes the mode and content of the communication. Many biologic systems, such as bees, sheep, and cattle, use gesture language, spatial relationships, sound, color, and pheromones to communicate. The reliability of implicit communication and display communication in trade-off space for unmanned platform applications has yet to be verified. However, no matter what the content of the communication between the unmanned platform or the unmanned platform and the central server is, robust network communication is the key to both strong coordination systems and most weak coordination systems.

5.7 Automation and autonomy

It is very interesting to make a difference between automation and autonomy, which people do not like to think about at ordinary times. They just use the words. They would not know that Westerners usually do not do like this. They often pay too much attention to the meaning of the basic concepts, and then the concept of a literal dig, and then the theoretical process of derivation and calculation or experimental practice verification analysis based on the former one, so the gap between Westerners and Easterners is often opened.

In an autonomous (self-constructed) system, when the interacting components act together, the combination rules are a superposition of fact and value, not only beyond (existing) mathematical calculation, but also beyond (existing) logical and nonlogical relations. And this will make each part to produce new properties and new features, new relationships, which are very different from the original ones. In contrast, automation favors predictable factual programming, while autonomy focuses on unpredictable value-based programming.

The world is made up of facts (relations), not of things (properties), and the process from things to facts is called organization. Put aside that it is difficult to find the building blocks of things, it would not be very useful even if we found it, because the secret of the world is the interaction between the building blocks, and we have never been able to say what these relationships are, even though we know they exist. Human beings are hyperharmonious logic (so they can tolerate, correct, and find errors). In the view of hyperharmonious logic, paradoxes do not necessarily need to be eliminated. Contradictions or paradoxes can be accommodated by a logical or theoretical system without causing the system to proliferate or making it meaningless. The hypercoordinated logic offers a new solution to paradox, arguing that we should accept paradox and learn to live with it.

In 1978, The Australian logician Priest published the paper "Surprise on Logic." The main difference between paradoxical logic and incongruous logic is that incongruous logic system itself is harmonious, while paradoxical logic itself is incongruous and contains contradictions. In his opinion, Godel's incompleteness theorem works only for systems that are coordinated; it does not work for systems that are incoherent and semantically closed.

Why are artificial intelligence systems far from what people expect? The screening case is that people have not yet found what the real relationship between the whole and the parts of the system is. Ashby pointed out that an organizing component will appear as long as one of the relations between A and B becomes the sacrifice of value or state C.

The biggest difference between automatic (including significance) selection and autonomous selection is that the results of automatic adaptive selection are almost certain (such as the output action of an automated machine), while the results of autonomous (value-related) choices are often uncertain (such as various contingencies and implications). Automatic production is generally factual reasoning, while heuristic intelligence is often value reasoning.

Human autonomy has a tendency to make up discrete state into continuous state and produce. This is probably because there's a part of the human brain dedicated to telling stories to yourself, to convincing you that everything is coherent and that the world has a cause and effect. Actually, it is not true in many cases. The cause and effect in the world are intertwined. The automation of causality is orderly, but autonomy, especially the wisdom of causality, is not necessarily orderly. Obtaining is not necessarily coveted truth, while losing could be the result of a beautiful eternal, such as a blessing in disguise.

Before we talk about automatic processes (mind), we will emphasize two kinds of processes: automatic (unconscious) processes and consciously controlled processes. There are a lot of automatic processes in human behaviors, and they extend into automatic percept, automatic pursuit of goals, continuous automatic evaluation of certain experiences, and so on. Here, we will discuss only two parts related to human-machine convergence: less effort that automatic processes require to perform (than nonautomatic processes do) and the low-level feedback required to cover automatic processes.

Automatic processes can be defined by at least three different parametric models: the behavioral mechanisms of guidance, the neuronal mechanisms involved, and the underlying process cognitive mechanisms. So, at the behavioral level, automatic processes can lead to rapid responses to stimuli; at the neuronal level, an amplified activity can be in a certain area of the brain; and at the cognitive level, this process does not need to be mediated by the attention system. The differences between the three models are important because behavioral and neuronal parameters lead us to a workable definition of the automatic processes currently occurring. This means that the user can predict the likelihood of triggering an automatic process under certain environmental conditions. We should pay particular attention to behavioral parameters, which can give users a better understanding of a particular type of interaction.

Different behavior parameters are used to define automatic processes. First, the automatic process is closely related to rapid reaction time, which measures the interval between the presence of a stimulus and the response (for example, pressing a button based on a screen display). Second, the automatic process is also related to the implementation of responsibility and obligation, that is, the inevitable implementation of the process. Third, there may be no interaction between the automatic process and other simultaneous processes; in other words, the performance of other processes will not be affected by the automatic process. Fourth, automatic processes are associated with high transitivity, so the performance level of automatic processes in different types of events remains constant. Fifth, automatic process is usually associated with unconsciousness; that is, the subject is often unconscious of the occurrence of events. Sixth, the automatic processes are insensitive to interference, so multiple stimuli do not affect the performance of the automated processes.

Many scholars believe that only the first two parameters can lead to the forced process of fast reaction time, although these parameters are often defined as separate values, such as whether the process is parallel fast or continuous slow. However, there is no consensus on the combination of parameters for measuring cognitive mechanisms. For example, many experiments are based on response time and stimulus duration to assess whether a subject's response to a stimulus is purely a bottom-up cognitive mechanism or not (i.e., being purely controlled by an external stimulus regardless of the subject's attention state). By adopting the research techniques of Treisan and Gelade in defining the Feature Integration Theory, many scholars believe that the process is bottom-up if the reaction time is relatively short and has nothing to do with the amount of interference. Similarly, for the duration of stimulus distraction, preattentional processing occurs when the stimulus is given to the subject for a short time (typically 200 ms), regardless of the amount of distraction. However, this definition is not clear enough about the process of comprehension and process intervals. As a description of automatic process, the learning-response process is quite different from the performance of preattentional process.

In general, factual interactions lead to automation, while value interactions can become autonomy. There are automation and autonomy in intelligence, and there are factual situation awareness and valuable situation awareness.

Automation, autonomy, and wisdom are the products of a human—machine (object) environmental system interaction. And wisdom is a kind of interaction at a higher level that goes beyond the first two, and it is a kind of value interaction beyond the fact.

CHAPTER 9

Reflection on human-machine hybrid intelligence

Contents

1. Thoughts on the human's intelligence in human-machine hybrid

High-speed traffic jams on narrow roads basically reflect the premise of classical economic fundamentals: the assumption that people act rationally. The basic premise of human-machine hybrid intelligence is very different from that of artificial intelligence: one is a real person, while the other is a dummy (virtual human). The so-called real person is the real person, while the so-called dummy specially refers to the ideal person, a hypothetical person.

A real person is very plump, with flesh and blood, having feelings and a sense of righteousness, needing food and clothes, being rational and emotional, having emotions, benevolence, righteousness, propriety and wisdom, so it is difficult to achieve human-machine hybrid intelligence.

If you think about it carefully, a dummy is not simple either. Although it is bloodless, it has to have a form; although it is heartless, it has to have a reason and an argument. In addition, it has to have concepts and definitions, premises and assumptions, formulas and codes, ethics and laws. So artificial intelligence is not simple.

Both real people and dummies are in the human-machine environment system. The people in human-machine hybrid intelligence involve thinkers, designers, manufacturers, users, sellers, maintainers, recyclers, and processors, etc., while the people in artificial intelligence also specifically involve many of the above personnel, but mainly include implementers and users.

Integrated Human-Machine Intelligence
ISBN 978-0-323-99562-7
https://doi.org/10.1016/B978-0-323-99562-7.00009-7

Whether real or fake, people are fallible. Intentional mistakes are planning a premeditation, and the mistakes that are not made intentionally are errors. Of course, there are some intentional and unintentional mistakes, such as quarreling while driving that may cause more serious accidents, which is called intentional error on laws.

Human's suanji is similar to intelligent calculation of quantum. The greatest advantage of them is the fast parallel calculation for a particular problem. Therefore, the greatest advantage of these computing forces can be applied to scenarios with low latency requirements. But human's suanji is stronger than the machine's calculation: it not only has good endurance and fast speed (powerful calculation ability and fast calculation speed) but also gets precise results (suitable to the situation). For example, the future intelligent command and control system of man-machine integration cannot only quickly calculate the real-time dynamic strategic planning system establishment of rush rivals, but it also can be combined with historical data, human's experience, all kinds of cases, all kinds of clues in the current game environment, timely implementation of the identification of false and fraud, and defense strategies of hiding truth and showing falsity.

Human intelligence can generate many elements according to the change of task environment, and then generate the corresponding basic unit, some of which could be quantified and broken down into basic AI units, like the periodic table of basic intelligence. The other part cannot be quantified statically but can only produce qualitative cross-domain analysis according to the changes of time, space, and logic in the specific task environment. This process can often produce creation or emotion.

Human intelligence benefits from creation and emotion, and is also subject to creation and emotion, which is due to the existence of occlusion relationship in different space-time logic and the human congenital defect, endless desire. Most successful human-machine hybrid intelligent systems can balance the two things well, forming a fusion optimization of sensibility and rationality.

Machines are still a long way from becoming part of the human body because intelligence, the richer form of human intelligence, cannot be fully represented by language. Many ideas cannot be formulated and programmed, such as empathy. Fundamentally, without addressing the mechanisms of human cognition, research in the field of intelligence may always be scratching its head, hardly touching the essence of intelligence.

2. The thinking about the current human-machine hybrid intelligence bottleneck

One of the bottlenecks of human-machine hybrid intelligence is that there is no physical theorem or law about what human-machine hybrid intelligence is. To understand this question, we must first explore the nature of intelligence. Different from a machine's intelligence, machine intelligence and artificial intelligence are fundamentally the conceptualization, systematization, and programmatic reflection of human intelligence, while fragmented knowledge and fragments of logic constitute the various complicated human intelligence. Fragmentation of knowledge + fragmentation of logic + implicit/explicit ethics and law constitute the human wisdom. The real intelligence needs combination, mixing, and fusion of different perspectives of logic from different areas. Real intelligence is the patchwork of hundreds of clothes, not the beautiful finished product.

After several highs and lows of artificial intelligence, rational authority and optimism behind formal methods is highly questioned. Symbols, connections, and behavioral technologies have not only brought progress, but they have brought more confusion about artificial intelligence. At the same time, genetic engineering and other transformative technologies are more likely to bring psychological, ethical, and thinking challenges to human beings themselves than in any other era of history. It is no longer important to use reason to find truth, but the meaning of existence is the core issue. The enlightenment replaced God with reason, but the result was the collapse of value, and existence was more and more nothingness. Heidegger tried to solve the problem of the nature of existence with *Existence and Time* and Husserl changed his thoughts from "Logical Studies" in the early stage to "Big Ideas" in the later stage. At the same time, the analytical philosophy leader Wittgenstein also turned his study from the early *Logic Philosophical Theory* to the thoughts of *Philosophical Studies* in the later period. These two transformations coincidentally dropped "logic" and moved toward "ideas" and "philosophy" respectively, which may not be a coincidence or accident.

"Logic" mainly involves judgment and reasoning, which belongs to consciousness activities at a higher level. To explain clearly the theory of judgment, it is necessary to conduct research on consciousness activities at a lower level, such as feeling and perception. In particular, it is necessary to conduct in-depth research on the mechanism of multiple representations of physical data and psychological information knowledge. The "name" in "nomenclature" mainly refers to the dynamic representation, naming, definition, and categorization. And the "Tao" reflects the mixing things of facts and values, things and relationships, including both objective "being" and subjective "should" involving the universal validity of both logic and mathematics, and covering the truth of the psychologic law of chance, which is a set of logic and nonlogic. Emotion maybe a complex of logic with other nonlogic.

Lao Tsu's "Tao" is extremely natural. For example, he said, "Wisdom comes with great falsehood" which means that the more one pursues wisdom, the more artificial things there are, and the more self-righteous elements there are. This requires turning our normal view of the world upside down and letting things see us, like an insect's compound eye. Cezanne said, "A good painter does not look from the outside but from the inside." Intelligent people also often use the opposite method to see the world, such as the old man who lost the horse, the spark on the Jinggang Mountain. We should not just use a scientific method to see the world. Intelligence and its philosophy should make people really open to the nature, so that the man-machine environment system in nature speaks to us and shows us in its own form. At present, the main problems of human-machine hybrid intelligence are as follows.

(1) elastic input of data, information, and knowledge; the representation of flexibility
(2) organic integration of axiomatic and nonaxiomatic reasoning; effective treatment
(3) the sink joint of liability judgment and risk-free decision; virtual and real complementary output
(4) synergy between human reflection and machine feedback; better adjustment
(5) cognitive balance between deep situation awareness and cross-domain resource management process; harmonious scheduling
(6) the generation of trust mechanism between man and machine
(7) the difference between machine knowledge and human knowledge
(8) interpretable threshold between human and machine
(9) the scope/content of machine lifelong learning is different from that of human learning

AI does not create new things. It can only do what people themselves know how to do. At present, as artificial intelligence is increasingly unable to meet people's expectations and appetites, the intelligent fusion brought by human-machine hybrid interaction is gradually coming to the front. However, human-machine hybrid intelligence, both domestic and foreign, both military and civilian, is not very satisfactory. So, what is the difficulty of human-machine hybrid intelligence?

As we all know, intelligence is not the product of the brain, but it is the ecologic product of the interaction of the human-machine environment system, including both objective data and subjective information and knowledge. If we say "beauty" is the product of the combination of subjective and objective, intelligence should also be the result of the fusion of subjective and objective, including all kinds of high, medium, and low elements, which is a dynamic organizational architecture that is both open and closed, diffused and converged. Intelligence is derived from values and meanings and generated by values triggered by facts. Data is a relatively objective existence, and only after being valued into information can it be condensed into knowledge in relevant situations, and then assimilated and adapted to many adaptive methods and effective means that can solve practical difficulties. And artificial intelligence is not a situation of real intelligence, nor a scene, but of the environment. More accurately, the man-machine environment "can be long and big" The situation and the scene are relatively small and have low penetrability, so it is difficult to produce a satisfactory solution and the optimal solution, which is why bad AI can have unexpected and explained reasons: it does not take the initiative in depicting scenes and situations. Now, the basic idea of artificial intelligence is to train a bunch of algorithms and then set the scene separately. In fact, it is difficult to drive real intelligence from data, information, and knowledge alone. Intelligence is a simple adaptive change from "being" to "should" (Hume's question). Intelligence produced by man is often intelligence (suanji), while intelligence produced by machine is often reason (calculation), and the intelligence generated by human-machine hybrid generally includes intelligence added to reason, which is a special suanji + calculation (simplified as jisuanji). In a sense, artificial intelligence is a kind of factual intelligence, while human intelligence is a kind of value intelligence. Of course, there are different degrees of truth and different degrees of value. Truth is often a mixture of facts of varying degrees and values of varying sizes. The advent of most contingency facts often comes from nonsituation, nonscene, and nonlogical value crossing.

The reason why people can grasp the direction is the dynamic value after analyzing the facts, so the subjective value (but not objective significance) facts of a thing or fact in a specific situation can be determined, so it is organic. But the machine is not subjective, and it is a calculation process without value and risk responsibility, and it is inorganic. Just like the usual choice, which is often as important as Mount Tai at the critical moment, value itself is not enough, which also needs to bear the consequences of courage and courage. Subjective judgment and emotional value are crucial, so the final decision to fire or engage cannot be left to a machine. Any thing or fact has many sides. To simplify, it can be called two-sided, such as yes and no, or 1 and 0. However, the fact of this thing will change its value with the development of correlation. No matter if it is fast or slow, short or long, human beings aptly embed the value of this thing and facts into an objective developing state, instead of labeling things like a machine. What is more interesting, this duality (or multifaceted nature) of the fact of things is naturally reflexive and will be triggered to be implemented in a specific task situation. "Yes" will become "no" and 0 may become 1, which is also one of the sources of uncertainty. Humans have ego, id, and superego and can constantly strengthen their subjective preferences (ego), while machines have no ego, id, nor superego and always perceive the world relatively objectively. Naturally, humans cannot easily reach a consensus with machines. Similarly, for a thing or a fact, the label of the machine will not be personalized elastic change, nor will it be extended to grow change, but people can. The label of machine data information knowledge will not grow, while the concept of human data information knowledge will change. For example, one inch of time is one inch gold, in which "one" and "inch" are changeable. The poem of "holding hands with you, I will grow old with you" is also very different from the original intention. People are very complex, but the machine can simplify them, Once labeled, one is fixed without strain and feelings of right and wrong. The way that humans make data not boring is to give them value to form information. For example, "3.15" is the Consumer's Day, 23 is Jordan, 658 is a district bus. The way to make information not boring is to condense meanings to generate knowledge, for example, $1 + 1 = 2$, an inch of time is an inch of gold. The only way to keep knowledge alive is to evolve into ever-growing intelligence.

A photon in classical physics has wave-particle duality, and a quantum in quantum physics has superposition state and entangled state. DNA in physiology consists of two antiparallel poly-nucleotide strands intertwined to form a double helix structure. In psychology, the relationship links of fact and value are also a kind of superposition, correction and spiral structure. All the analogy with the appearance of some sort of "divine" penetration and essence, the analogy of physics, physiology, and psychology is no exception. Actually, intelligence is essentially the mutual influence, interaction among the psychology (consciousness), biology (nerve), physics (environment), product of simple call character (machine) environment system. Generally speaking, artificial intelligence uses symbols, behaviors and connection to carry out formal representation, reasoning, and calculation of objective facts, rarely involving the judgment and decision of value-based and responsibility-based causality. Whereas, depth in deep situation awareness refers to the integration of facts, values, and responsibilities (i.e., the complementation of real and virtual). Situation and state refer to the objective facts of data and information and an objective part of knowledge (such as dominant and parameters of time and space, etc.), which are simply called the chain, while sense and percept involve the parameters of the subjective value (such as the bright hope, effort, etc.), which are called the value chain. The depth of the situation awareness is the fact that chain and value chain are intertwined in the double helix structure, which can implement valid judgment and accurate decision-making function. In addition, humans focus on the subjective value control calculator, while machines focus on to the objective fact process calculation, which is also a "double helix" structure. How to achieve the appropriate matching of the base pairs (space-time) between the two "double-feeding" structures is still a problem that has not been solved by all countries, so how do we characterize these parameters? How do we build this model? According to psychologist Kahneman, people have two egos: the experiencing ego and the remembering ego. The experiencing ego is responsible for actions and decisions, and the remembering ego is responsible for interpretation and reflection. Similarly, a human has two kinds of intelligence: factual intelligence and value intelligence. Factual intelligence is responsible for objective and rational intelligence, while value intelligence is responsible for subjective and perceptual intelligence. In short, true intelligence does not happen on your phone, it happens in your life.

Humans will flexibly use what they have learned. Their wisdom is living, full of changes and dialectics, both combining and separating, both spreading and gathering, handling particular cases in particular situations. Humans may not need to perfect the definition of the line about when, where, or how to balance and consider all aspects. For example, the white horse is not horse in which there are facts, value, and responsibility, or different combinations of these changes. (In the input end, it is a fact that a white horse is not a horse. But that the white horse is not horse is not a fact, because the white horse is indeed a horse, which is a matter of value. That a white horse is not a horse is an important symbol of human intelligence, which reflects a hard characteristic of machine intelligence to characterize: fact reflexivity representation). Machines only have data and formulas, and machines have no purpose. All human behaviors have a purpose, which is value and responsibility. Purpose can be divided into far, middle, and near purpose, while responsibility can be large, medium, or small. Human autonomy and denial often involve responsibility and value, not just facts. Further, a concept can be judged by three axes. One is the fact axis, involving time, space, attributes, physics, logic, and other objective reality. Another is the value axis, which involves the subjective and possible aspects of individual psychology, art, relationship, ethics, and nonlogic. And the other is the responsibility axis, which involves the general psychology, art, relationship, ethics, nonlogic, and the other three possible aspects. Its connotation and denotation often change in the coordinate system decided by these three axes, thus forming a bright and colorful intelligent world and dazzling meaning (to) form (type) situation.

Classification is a basic way for human beings to understand the world. The division of things is the starting point of conception and the premise of all thinking. The science and technology rest department is established from the starting point of classification. But people only pay attention to the fact classification, and they ignore the value and responsibility classification, especially the mixed classification of the three. In addition to the signifier and the signified, the concept also has a kind of kinetic signifier, that is, a kind of random dynamic uncertain pointing, just as the children said that karma, such as "ice cream mood" "unhappy happiness," which can make the boundary between concepts become optional at any time and can be arbitrarily crossed.

The "*The Book of Changes*" is dialectics: "zhiji" (signs) is the universal connection, "qu" (seizing the opportunity) is the unity of opposites, and "biantong" (flexibility) is the change and development. Tao is the natural order of two sides. In real interaction, there are often state collision, situation collision, sensory collision, and percept collision. For the deep situation awareness of human-machine hybrid intelligence, situation awareness, fact, value, and responsibility are dynamically linked together. From the stages of observation and representation, adjustment and reasoning to judgment and decision-making and implementation of actions, human intelligence is not only full of counterfactual, but also mixed with a lot of antivalue and anti-responsibility. At the same time, human analogy can solve various nonmapping relations that cannot be solved by machines. In deep situation awareness, trend is direction and speed (direction is more important); state is degree and magnitude (degree is slightly ahead); percept is essence and connection (essence is particularly prominent); sense is phenomenon and attribute (phenomenon is greater than attribute); depth is the integration and interaction of the human-machine hybrid environment, the integration of the interaction. The difficulty of situation awareness lies in the promiscuous nature of state and potential and the promiscuous nature of perception and knowledge. What is more difficult is the mixing of state and trend, the mixing of sense and percept, where there are true and false, truth mixing with false, virtuality and reality, and virtuality combining with reality, and so on. The standard for evaluating the quality of deep situation awareness is the ability of being a person, machine, loop, situation awareness, and auxiliary lines. If you do well, you will be able to solve it easily. If you do not do well, you will be able to go in the opposite direction.

Between man and machine, the difference between autonomous intelligence and others-oriented intelligence is assimilation and adaptation on the surface, but actually it is the conversion degree and efficiency of assimilation and adaptation. Autonomy is simply a "should" issue, focusing on a value issue. When studying child psychology, Piaget believed that the development of children is constantly happening under the interaction with the external environment. Child development is not simply a process of constant external stimulation; in fact, it must rely on the internal structure of the child. Children's activities and external stimuli are equally important. With the growth of children's age, their cognitive development involves four key words: assimilation, adaptation, balance, and schema.

Children's assimilation, adaptation, balance, and schema are based on objective facts at the beginning, such as physiologic needs of eating and drinking, puling and sleeping, etc. With continuous growth, assimilation, adaptation, balance, and schemata of value are gradually formed, such as love, hate, and worry. Later, the assimilation, adaptation, balance, and schema of responsibility are derived, such as ethics, integrity, shame, and courage.

(1) Assimilation refers to the process of filtering or changing stimulus input by learning individuals. That is to say, when the individual feels the stimulus, he puts it into the original schema in the mind and makes it a part of himself, to strengthen and enrich the action of the subject.

For example, I used to cook fish in a pan, which cost me 20 min. Now I have bought some beef, I think of pot cooking naturally. If successful, it is the assimilation of skills.

(2) Adaptation means that when the external environment changes and the original cognitive structure cannot assimilate with the process of reorganization and transformation of children's cognitive structure caused by the information provided by the new environment, the individual's cognitive structure is changed by the influence of external stimuli. It is a process that individuals change their actions in response to objective changes.

For example, there is no meat in the market now, but there are only vegetables. I also spend 20 min cooking them, but the cooked dishes taste bad. So I decide to cook it for 10 min. This is the procedure I conform to the way the vegetable is processed.

(3) Balance refers to the process in which learners continuously make their cognitive development transition from one balanced state to another by means of assimilation and adaptation through a self-regulation mechanism.

For example, that I could cooked meat was in equilibrium. Then a new dish appears, and I finally learned how to cook it after I was not able to cook it, so I got to a new balanced state. And this process is the process of equilibrium.

(4) Schema is a kind of structure and organization, which will cause migration or generalization due to repetition in the same or similar environment. The first basic abilities are inherited from birth, but it keeps changing, as in "*The Book of Changes*" and developing later in the process of adapting to the environment, forming cognitive schemata (structures) of a different nature.

For example, I can cook meat and vegetables and can cook millet porridge. I also published a book of food recipes, "*A Bite of China*", based on experience.

This is an interactive and constantly changing development process. The development of the child and the new man-machine system is full of assimilation, adaptation, and balance, which is the process of continuous change and development of the schema system.

The same is true between man and machine. Even newborn babies and new human-machine systems have their own simple schema systems. We cannot ignore the child or the original human-machine schema system and blindly emphasize the external environment for children to learn.

Only by making the child and the new calculation system experience it all by themselves and emphasizing the child's and the new calculation system's own movements and activities will it be effective, such as grabbing objects by themselves. Only when the child and the new man-machine system rely on the existing structure, namely the schema system (which refers to the child and the new man-machine system action, constantly grasping objects), can we achieve the process of assimilation.

On this basis, the child and the new man-machine system are more likely to try other objects, which is a generalization of assimilation. And the child's own internal structure changes because of this process to adapt to reality, which is adaptation.

DARPA'S "Dark green" accusation system does not do well in this regard, so it does not get the expected application effect. The mechanism of intelligent human-machine hybrid intelligence and situation awareness revealed by the entire system is relatively vague and confusing. The resulting intelligence only describes factual computers, and it lacks human emotion, value, and responsibility. The assimilation and adaptation are unbalanced, and the direction and process of development between the human and machine schema system are constantly changing.

The essential problem of any human-machine system disharmony lies in how to grasp "changing" and "good" rather than "fast" and "acting" Otherwise, humans are not humans, machines are not machines, and the environment is not the environment. Their advantages have not been released respectively. They remain the same when they should change, and they change when they should not change randomly. In addition, the methods, timing, and functions of the human-machine hybrid should be appropriately "good" Only when they are not early, not late, not fast or slow, can they shine their respective advantages and achieve optimal matching. In a real open environment, the resulting degree of intelligence and active effectiveness can be maximized.

One trend of human-machine hybrid intelligence is "hardwarizing" software and "softwarizing" hardware, the humanization of machine parts and the continuous mechanization of humans. In fact, it is not necessarily a good phenomenon. Humans should do human things, and machines do machine things. The core of its hybrid needs to emphasize that there must be a range in all human-machine hybrids. Any intelligence is not what is always intelligent anywhere at any time. It always has limitations, and so do humans themselves. So how to find that range is very important (when you take an overall view of the current disciplines, it is not convincing that we can study the "just right" in human-machine hybrid intelligence by using the existing mathematics, physics, physiology, psychology management, and so on).

In natural science, people often use mathematical equations to describe phenomena. For example, if time (T) is taken as a variable, the change of cognitive operation is equal to the function of body state (S) and response to external stimulation (R), where S refers to the physiologic and psychological state of the body, and the memory in the brain and so on. When the external stimulus acts on the body in a specific state, it will produce results and changes, that is

$$T' - T + I, \hat{x} = f(S,R),$$

where T' is the changed time, T is the current time, and I is the increased time.

Cognitive science believes that the working principle of computers is the same. In a short period of time, the memory stored by computers is equivalent to the state of the body; calculation input is equivalent to some kind of stimulation applied to the body. When a certain input is given to the calculation, the calculation starts the operation process, and its internal changes take place to obtain the results. The operation process of a calculation can be regarded as the change of its state in each unit of time. Calculation programs can be used to simulate people's strategic level, calculation language can be used to simulate people's primary information processing, and calculation hardware can be used to simulate people's physiologic processes (activities of the central nervous system, neurons, and brain).

In fact, a human's psychological structure is different from physiologic and physical structure. It is not only subject to the body itself but also the result of adapting to the environment. Therefore, in $T' - T + I, \hat{x} = f(S,R)$ and $\hat{x} = f(S,R)$, a human's R is different from machine's R. The human's R involves not only external stimulus r, but also internal stimulus r, so it is $\hat{x} = f(S,r,r',r'')$. In the deep situation awareness computing + suanji system, the external stimulus r can be regarded as the internal state stimulus, the internal stimulus r' can be regarded as the internal state stimulus, and the internal stimulus r'' can be seen as a trend stimulus. Perhaps, sense can be seen as state stimulus, and percept can be seen as trend stimulus.

At present, in the intelligent platform of human-machine hybrid, humans are necessary in the system environment (but we have to ensure that people are normal people). Human-in-the-loop is the direct part of a system, both monitoring and controlling. Human-on-the-loop is the indirect part of a system, mainly supervising and controlling. An intelligence system of human-outside-the-loop is out of control. Then how do we study a human-machine environment system and make it an engineer? First, we should study the human beings, including humans' senses and perceptions. Secondly, we should study the machine and how to transfer these perceptual functions to the machine (equipment) and mechanism (management). Thirdly, we would study the environment, including state and trend generated in various environments (hereinafter referred to as situation). The research on the three is not necessarily sequential, but it can also be reversed, inserted, mixed, and integrated. People's percept ability was studied in the past, while machines (equipment) and mechanism (management) are studied nowadays with the development of artificial intelligence. The future development trend is to study how to combine these two factors. That is the problem of deep situation awareness in human-machine hybrid intelligence, which is also the key problem in the study of human-in-the-loop.

Just as people often start to understand the world from witchcraft and mythology, cognitive science starts from a "wrong" analogy: computers are not like people at all. The so-called symbol is just a pattern. Any pattern is a symbol as long as it can be distinguished from other patterns. Computers, whether electronic or quantum, are artificially defined and agreed "physical symbol systems" That is, they emphasize that their research object is a specific material system. For human beings, their research and applications are flexible and personalized "psychological symbol systems" whose research object is an abstract value meaning system.

For learning, people often mistakenly think that a human's learning is standardized and regular. In fact, it is a misunderstanding that a human's real learning is not completely integrated and systematic, but permeated with a large number of personalized and flexible metaphors and analogies in the learning process. The shattered knowledge and method skillfully run through and adhere to the so-called standardized knowledge learning process. As a result, educators consciously or unconsciously use the way of reverse order to tell students that the knowledge in this subject and its books is systematic and complete. The implicit cognitive methods behind the discovery process of acquiring this systematic knowledge, however, are ignored and omitted. And these unexplained parts are just real learning. At the same time, it is also beyond the reach of machine learning. In contrast, a human's learning may be nonrepresentational or weak representational, which is a kind of understanding learning, while a machine's learning is a kind of "nonunderstanding" representational learning. Learning can make people better and more calmly upgraded and deal with the unknown things, but this is not always the case with machines. In addition to the state (such as action sequence, the number of words and numbers, etc.), there is also the trend (development and change trend) in a human's learning with the chain effect of looking down. In a machine's learning, no matter if it is deep learning, reinforced learning, or other kinds of learning, there is less override from trend to state, but only following the steps from state to trend. During this process, many tentative stimuli to choose and adjust are gone. For human learning, even the expression (word) of the same concept often has different objective and subjective components. How we can reach a consensus as far as possible may involve the ratio of fact and value. The language of interpersonal communication is a compound two-way channel mixed with signifier and signified, while the current human-machine interaction can only point to a single channel, which leads to the fact that there are no overtones in the current intelligent communication. Perhaps in the near future, human-machine intelligent communication will form a compromise interaction mode of "signifier + signified" between signifier and signified to facilitate the development of intelligent communication systems between contact and machine. Machine learning is often in name only, while human learning can be both in name and in reality, and they can also be in reality but not in name, neither in name nor in reality. It is the value domain that breaks the time-space domain of facts, leading the value domain trend, thus forming a distinct, proportional, and decision-making chain. Human wisdom is manifested in dynamic representation + dynamic reasoning + dynamic planning + dynamic implementation, in which the dynamic is based not only on facts and values but also on responsibilities and obligations. But machine intelligence is not like this.

The key to the high-speed development of the intelligent platforms in the future is the coordination of human-machine environment systems, where the humans include designers, manufacturers, managers, operators, consumers, maintainers, etc. The machine here not only refers to the software and hardware in the intelligent equipment, but it also involves the street connection mechanism between various links in the industrial chain. And the environment here involves a multifield "political industry—university—research business" cooperation and collaborative environment. Through the interaction of state, trend, sense, and percept among humans, machine, and environment is realized the application of new AI+, such as precise force, data integration, and so on.

In conclusion, it is better to start studying complexity with simplicity, it is better to start studying facts with value, it is better to start studying group from a single point of view, and it is better to start studying organization with internet.

It is better to start studying characterization with gathering, it is better to start studying decisions with personality, it is better to start studying human-machine relation with border, it is better to start studying the unknown with the known, it is better to start studying model with an analogy, it is better to start studying content with form, and it is better to start studying consciousness with substance.

It is better to start studying intelligence with distinguishing, it is better to start studying commanding with controlling, it is better to start studying algorithms with data, and it is better to start studying beginning with the result.

Vice versa.

How to realize the organic combination of deep situation awareness function and capability in human-machine hybrid intelligence is the main indicator to measure the quality of the system. The capability is mainly to generate intentions, while functions focus on realizing intention. One of them is active and the other is passive. The intention is not produced by a blink of an eye but the interaction between humans, machines, and the environment. Therefore, this is a complex and mingled problem, which may involve some basic problems of division of labor and coordination. The best way to study complexity is to start from simplicity. For example, the best "learning" time of humans and machines is childhood. One reason why children learn language relatively quickly is their fast brain development. In addition, another important reason is that the formation of all their concepts and knowledge is related to the image of objective facts. Rarely thinking, they form value significance for no reason. "Making mistakes actively" may be an important way to obtain creative thinking ability. In addition to slower brain development, adults' difficulties in learning language are often related to social customs, personal habits, environmental constraints, and boundary conditions. Adults are always afraid, unwilling to make mistakes, and they always hesitate before they act. This preconceived causal way often restricts the outbreak of various relationships.

In essence, a human-machine hybrid is a form and method of fact and value changing. It can never and ever be static. The process of "constantly innovating the ergonomic structure from the inside, that is, constantly destroying the old and creating new structures" in the human-machine hybrid is just like a biologic term "structure mutation" The human pattern recognition is fundamentally different from machine pattern recognition. Human pattern is not only a state but "complacency" of trend as well. If machine pattern recognition is a factual entity, human pattern recognition is a virtual entity mixed with fact and value.

So how to establish the trust relationship between developers, users, and the system will become more and more critical. At present, there are two solutions for a human-machine hybrid. The first is to allow people to participate in the training process of the system (this involves the problem of when, where, and how people effectively participate in the system). The second is to allocate as many decision-making tasks to people as possible (this involves how to select the decision-making tasks suitable for people and prevent the degree problem of "projection effect"). Although experiments have shown that these two points can increase people's trust in machine systems, whether this conclusion is extensive needs more in-depth analysis and research.

Human intelligence can be expressed as dynamic representation + dynamic reasoning + dynamic planning + dynamic implementation.

What is the concept of intelligence? This key problem is not clear yet. Now the combination or "hybrid" of human and machine can only be improved within the range of artificial intelligence. Now artificial intelligence is based on statistics and optimization, but human intelligence is not based on statistics. Artificial intelligence learning needs a database, but people do not use a database. If biologic systems are selected according to the optimization law, there is only one optimal species in the world. However, if people are selected according to the optimization thinking, there will be a situation where a hundred flowers bloom and a hundred schools of thought contend. It is not time we discuss this issue.

3. Thinking in the field of human-machine hybrid intelligence in the future

The evolution of human civilization can be roughly divided into Western civilization and Eastern civilization, and human's understanding of the field of intelligence can also be roughly divided into two systems: the East and the West. The development of the field of artificial intelligence mainly extends the scientific and technologic context of Western civilization (logic + experiment). As a reflection of the field of human intelligence, Eastern civilization also plays an important role (insight + balance). It can also be considered that the Western focus is on logic and algorithm while the Eastern is more nonlogic and nonarithmetic.

The future intelligent form of human-machine hybrid needs to organically integrate the reasonable parts of the East and the West to form a new intelligent matching mechanism. This matching mechanism includes two parts, mutual correspondence and mutual cooperation. The machine is regarded as an object based on deterministic data, algorithms, and calculation power, so humans should be an object based on stochastic knowledge, computing theory, and calculation, in which the knowledge has the characteristics of subjectivity, strong convergence, bouncing, and uncertainty.

In a sense, intelligence is the process of finding the best alternative. The alternative here includes substitutes, alternative schemes, alternative systems, etc. The search is a mixing processing of calculation and suanji, where suanji often involves macro direction and inner truth, and algorithms are often associated with specific processes. Human suanji involves explicit and implicit knowledge, focusing on the integration of value and fact. Human and machine computing includes describing programmable knowledge, focusing on fact. In a machine's calculation, there is both opposition and unity, but in a human's suanji, harmony can often be valued.

Both artificial intelligence and human intelligence have a common disadvantage: they can easily hurt themselves. That is, intelligence is mistaken by intelligence. Therefore, in the processing of data, information, and knowledge of human-machine hybrid intelligence, it is necessary to establish a predictable responsibility allocation mechanism and clarify whether, when, and to what extent to use what algorithm system. Therefore, there should be both technology and art in the future human-machine hybrid intelligence. In other words, concept, definition, reasoning, and decision-making of the intelligence involving human-machine hybrid is not fixed. There is an intermediate area between state and trend—situation area—where there are both state and trend, both facts and value, both data and information knowledge. In the situation area, there are both axioms and nonaxioms, intuition (nonlogic) and logic, reflection and feedback (reflection is a dynamic virtual real compound feedback).

At present, human-machine hybrid interaction lacks dynamics, and the qualitative analysis between them has not been completed, so it is quite difficult to quantify it. For example, how do we make the machine "understand" the change of human intention in different stages? How do we make people understand various calculation results of the machine? Sometimes, a large and comprehensive database and knowledge base may be a big obstacle because many changing factors are difficult to be (or cannot be) expressed by parameters. For example, a baby's cry can be caused by hunger, pain, illness, or sleepiness, and it may also be caused by the above comprehensive factors, but it is difficult to use a fixed database or knowledge base (even conventional knowledge graph) to characterize. The reward and punishment mechanism in machine reinforcement learning is very different from that of human beings. Human rewards and punishments have "righteousness" (value) in addition to "benefit" (fact). Similarly, the mechanism of the state, trend, sense, and percept of machines is quite different from that of human beings. Machines are basically "convincing people with reason" while human beings are "blending of emotion and reason" The interactions between machines and humans are both one-way, but the communication of humans toward machines is multidirectional loops of human-machine, machine-machine, human-environment, machine-environment, and human-machine-environment between human-machine environmental system, among which there are not only a large number of "interconnections" but also more "mutuality" People are the active part of the environment, and machines are only artificial passive tools. For example, the interfaces of many machines (such as various reminding methods of mobile phones) will not adapt to the changes in the environment, tasks, and people.

Human beings generally associate and judge things through daily common sense, and some of their complex reasoning is also related to dynamic expectation, while machines are connected and analyzed by incomplete data and non-(human) common sense, and there is no human-like expectation mechanism. Fundamentally speaking, the smartness and narrowness of machines are different from those of humans. The model of human problem processing is a creative cognitive computing model that continuously integrates across domains in an infinitely open and nonlinear environment, while the model of machine problem processing is an empirical computing algorithm model in a limited, closed and linear environment. At present, for all important human-machine systems, the final judgment still counts on humans because the essence of these problems not only belongs to scientific and technologic problems but also involves a large number of nonscientific and technologic problems such as environmental noise, social humanities, ethics, law, and so on. Artificial (machine) intelligence is a fact algorithm where people use logic to write, considering the combination of rules, such as eating with chopsticks or knives and forks. But human beings use nonlogic (higher-order logic mixed with facts and emotions) to carry out dynamic value algorithms, which are more appropriate responses. For example, humans can hold chopsticks or knives and forks, and they can also eat with their feet or other tools. Artificial intelligence is a "yes or no" issue, while human intelligence is a "should or should not" issue. Function is the passive realization of a tool's non-adaptability, and capability is the active realization of life adaptability. People do not know themselves, especially the real process and essence of the formation of people's cognition and feeling. The neural network of humans is not an artificial neural network but a multimodal biologic organization intertwined in three dimensions. Human beings are environmental. Few people are unaware of the snow in summer. Whether they can be aware and conscious of their own behaviors is often an important difference between humans and machines.

Real intelligence is not absolutely correct at the beginning, or it may make directional mistakes at the beginning, but it can better represent the size and quality of intelligence by constantly adjusting in real time in the process and properly adjusting the program, timing, and mode in the process. As Heisenberg said, "Any understanding must ultimately be based on natural language, because only there can we really touch reality." In fact, the language of a child is different from that of an adult. The same concept or statement used by adults has some discovery and exploration in children's languages where there is more emotion than knowledge, more value than fact, and more virtuality than reality. When he or she ponders this concept or statement, he or she always finds the best way between repeatable and unrepeatable approaches to reach self-consensus, and it can accurately migrate to other situational tasks in the future. Perhaps it is more accurate to change the first and second sentences of Wittgenstein's *Tractatus Logico-Philosophicus* respectively into "the world is everything that happens and does not happen" and "the world is composed of facts and values, not things."

A friend of mine (New York Old Bear) suggests, "In fact, any system which is big enough to a certain extent will have problems of interpretability. This is especially true for deep learning because no one knows how a huge number of parameters work. For other systems, for example, if a reasoning system is large enough, its behavior is difficult to be accurately known. However, it can be known in principle, regardless of the cost. This is in contrast to (machine) deep learning." The core of semantics lies in value. The greatest difficulty of interpretability lies in the understanding and explanation of semantics. Learning is to establish the connection of facts, and understanding is to realize the connection of values. While reconstructing each other, there is also a huge gap between facts and values.

Different from machine learning, human learning is the connection between compound facts and values. At present, whether to create a new evolvable machine learning model is the touchstone to measure whether it is a new generation of artificial intelligence. Nowadays, machine learning cannot be ruled by one single algorithm but must be composed of various mathematical models. According to different applications, the most suitable machine learning model will be selected. Of course, machine learning must be adaptable to the scope of application, from multidomain applications to single domain applications. In the current algorithm field, it is impossible to produce any model with more powerful computing power than human-machine hybrid learning. A set of calculation systems or suanji systems of human-machine hybrid may better represent the development trend of the intelligent field in the future.

With a new round of scientific and technologic revolution developing, especially the breakthrough of network communication technology and the strengthening of artificial intelligence technology, the field of human-machine hybrid has also entered a new era. In the current era, the content and form of the human-machine environment system relationship are very different from those in the past, and the selection of the human-machine hybrid strategy and the effect of interaction strategy are different from those in the past. In this case, understanding the current human-machine hybrid intelligence with the traditional concepts and values of human-machine interaction is likely to make this aspect of research fall into a passive situation. Therefore, we need to break through traditional thinking such as fact and value analysis to understand the current intelligent problems and relations of the human-machine hybrid. Any intelligence is a new solution to specific problems. However, when the original problems are solved, new problems will inevitably arise. Therefore, new intelligence is needed to solve new problems, which determines that human-machine hybrid intelligence is always in progress and not completed.

4. A revolution is sweeping artificial intelligence

Anything has many aspects, and its advantages often become its disadvantages in a certain situation. Under other conditions, the disadvantages can also be transformed into advantages. For example, the power of something often makes its fatal hidden danger ignored, resulting in the continuous weakening of the future. On the contrary, a weak thing sometimes strengthens its hidden weakness and becomes stronger and stronger. Think about it carefully. Many things in the world are as changeable as the yin-yang fish in the Taiji Bagua (Eight Diagrams).

Taiji Eight diagrams is a circle at the beginning. It is the same inside and outside, which is in vain. But when people give a concept, it becomes a real boundary. In the beginning, the circle was integrated, without the distinction of the black and white and the line in the middle. The concept was artificially expressed in three forms: white was yang and black was yin, and the line in the middle was the unity of black and white and finally belongs to black and white. Someone regards the "*The Book of Changes*" as a decision-making system and an effective, simple model. Yi, the first Chinese character of the name of "*The Book of Changes*" is firstly explained as "easy" that is, greatness is in simplicity. The second meaning of Yi is "change" which means it should not be confined to form. Yin and yang are embracing, not distinguishing. Everything is a dynamic balance. Yin and yang will not be transformed. Yin and yang are not antagonistic or contradictory. They are two aspects of one thing.

In classical physics or the "Newton system" it is generally believed that everything was destined from the big bang. The evolution of the world is explained by some mathematical formulas, which are expanded from the initial conditions in the most accurate way to describe the world. To this end, physicists use the language of classical mathematics to express these initial conditions with real numbers. Nicolas Gisin, a physicist from the University of Geneva in Switzerland, said, "These numbers are characterized by infinite decimal places after the decimal point, which means that they contain an unlimited amount of information."

There are many typical real numbers, which are composed of a series of completely random decimals. The most well-known π is only one of them. People rarely need them in daily life, but their existence is a recognized assumption in classical mathematics. As a result, these numbers appear in many formulas in physics.

However, here is a question: since our world is limited, how can it contain unlimited numbers and numbers with an unlimited amount of information? To avoid the contradiction of "finiteness containing infiniteness" Professor Nicholas Gisin suggested returning to the source of classical physics: changing the mathematical language so that we do not have to rely on real numbers.

Perhaps human beings can always find a balance between possibility and impossibility, "should" and "should not," existence and nonexistence, consciousness and nonconsciousness, fact and nonfact, value and nonvalue, family and nonfamily, to realize the intention and purpose of change. In human-machine hybrid intelligence, perhaps, people's situation awareness or deep high-order situation awareness is very different from the equation in classical physics, which accurately describes the evolution of the world determined by the initial conditions of the big bang, that is, people's daily experience and intuition often contradict this deterministic view: is everything really written in advance? Is randomness just an illusion? Nicholas Gisin has been analyzing the classical mathematical language used in modern physics. He found that there was a contradiction between the formula explaining the phenomena around people and the limited world. He suggested modifying the mathematical language to make randomness and uncertainty a part of classical physics, to make it closer to quantum physics. So, should we also regard randomness and uncertainty as a part of the research of situation awareness in traditional human-machine hybrid intelligence to make it closer to reality?

Another difference between the two mathematical languages, classical mathematics and intuitionistic mathematics, is the authenticity of propositions. In classical mathematics, according to the law of exclusion, a proposition is either true or false. But in intuitionistic mathematics, a proposition is either true, false, or uncertain. Therefore, uncertainty is acceptable in intuitionistic mathematics. Compared with the absolute determinism advocated by classical physics, this uncertainty is closer to people's daily experience. Similarly, in human intelligence and human-machine hybrid intelligence research, this uncertainty is also an important part that is closer to reality, which is often intentionally or unintentionally ignored. For this, we also suggest modifying the mathematical language and intelligence research ideas (especially human-machine hybrid intelligence) to make randomness and uncertainty a part of classical intelligence, to make it closer to the research of real intelligence.

Therefore, randomness also exists in deep situation awareness in human-machine hybrid intelligence and quantum physics. Professor Nicholas Gisin said, "Some people try to avoid randomness at all costs and include other variables based on real numbers. But in my opinion, we should not try to make quantum physics closer to classical physics by eliminating randomness. On the contrary, we must eventually bring classical physics closer to quantum physics by introducing uncertainty." These sentences also apply to the field of intelligence research. Then we need to find out in theory where the boundary between human and machine is, how facts and values are mixed, and how intuition and logic cooperate.

People's view of the world is constructed through the language they say. If we choose the language of classical mathematics, it will be easy for us to think around determinism. On the contrary, if we choose the language of intuitionistic mathematics, it will easily for us to tend to uncertainty. In the field of artificial intelligence, algorithmists and programmers who dare not face intuition are everywhere.

Professor Nicholas Gisin explained, "I now think that we have accepted too many assumptions of classical physics, which means that we have integrated a determinism that may be unreasonable. On the other hand, the classical physics will also become uncertain, just like quantum physics, if it is based on intuitionistic mathematics, and will be closer to our practical experience, opening up other possibilities for our future."

Perhaps, with the continuous development of research in the field of intelligence or human-machine hybrid intelligence, this change may constantly change the current research results, but it will make it easier for people to understand the essence and connotation of intelligence: the existing mathematics, computers, automation, and programming cannot meet the realization side, and finally give up the world outlook that "everything is destined" making room for new perspectives, randomness, opportunities, and creativity.

5. Artificial intelligence and ethical questions

The rapid development of artificial intelligence has also brought some troubles and uneasiness to people's life. Especially after the singularity theory was put forward, many people believed that the rapid development of machines would bring great danger to humankind. The following machine accidents and the generation of machine weapons further confirmed people's impression. Therefore, the research on machine ethics and machine morality is emerging one after another. This part will start with the ethical issues of artificial intelligence, first discussing the ethics of artificial intelligence and its related concepts, and then some focused issues of artificial intelligence ethics, such as whether artificial intelligence can replace human beings and the responsibility of artificial intelligence, and finally giving some views of the author.

The English word "ethics" comes from "ethos" in Greek, whose meaning is similar to that of "mores" in Latin which means customs and habits. There are many schools of ethics in the West, including Schopenhauer's voluntarism ethics, James's pragmatism ethics, Spencer's evolutionary ethics, and Heidegger's existential ethics. Among them, existentialism is the most influential ethical school in the West. It always takes freedom as the core of its ethics and believes that "freedom is the only source of value."

In China, the concept of ethics dates back to the sixth century BC, and there were single Chinese characters of "lun" (ethics) and "li" (morality) in *The Book of Changes* and *The Book of Odes*. The former refers to people's relationships. The ethics in the "three cardinal principles and five constant principles" and "ethical cardinal principles" is human relation laws. The morality refers to order and reason, and it refers to the code of conduct that people should follow. Similar to the West, the meanings of ethics are very different among different schools. Confucianism emphasizes benevolence, filial piety, fraternity, loyalty, faith, and moral cultivation. Mohism believes in "love and mutual benefit" while legalism pays more attention to the rule of law than education. Human nature is evil and depends on the law to restrict each other.

Generally speaking, ethics is a branch of philosophy, and it is a science that studies social and moral phenomena and their laws. It is necessary to study it, because ethics can not only establish a relationship between people but also restrict and affect people's behavior through underlying values. It is hard to imagine whether people's society would have human relations and order or not without the concept of ethics.

In fact, many scholars have started to study the relationship between machines and humans and expressed their opinions before the term "artificial intelligence ethics" was promoted. As early as 1950, Wiener worried that automation technology would cause "the depreciation of the human brain" in his book *The Human Use of Human Beings: Cybernetics and Society*. In the 1970s, Dreyfus successively published articles "Alchemy and Artificial Intelligence" and "What computers can't do," coming to the conclusion that artificial intelligence will fail at the level of biology and psychology. The concept of machine ethics (similar to artificial intelligence ethics) comes from the article "Toward machine ethics." This article clearly puts forward that machine ethics focuses on the behavior results brought by machines to human users and other machines. Anderson, one of the authors of the article, suggested that they should also afford some social responsibilities and have ethical concepts as machines become more and more intelligent. This can help human beings, and machines themselves make better intelligent decisions. Coincidentally, in 2008, Professor Noel Sharkey, a British computer expert, once called on mankind to formulate moral and ethical standards related to machines (robots) as soon as possible. At present, there are relatively many studies on artificial intelligence ethics abroad, such as "The Road Map of Robot Ethics" issued by the European robot research network in 2005, "The Charter of Robot ethics" issued by the Ministry of Industry, Commerce and Energy of Korea, and "The Robot ethics" sponsored by NASA (note: we believe that artificial intelligence ethics is not much different from machine [robots] ethics in essence, and the two can be replaced by each other). Moreover, the relevant foreign literature is also relatively rich, mainly focusing on robot law, safety, and social ethics.

Domestic researchers started related research late, and the research is not as systematic and comprehensive as foreign research. However, in recent years, scholars have also focused on the ethics of artificial intelligence. Relevant documents include "*Ethical boundary of robot technology*" "*Human rights: can robots get it?*" "*Do we want to give robots*" "*human rights?*" "*Make rules for robots, hurry up?*" "*Preliminary exploration of artificial intelligence and legal issues*" etc. It is worth mentioning that from the above literature, it can be seen that Chinese scholars have changed from simple technical ethics to ethical research in human-machine hybrid interaction, which is undoubtedly a great progress.

However, it is a pity that there are still few formed laws and regulations to restrict AI technologies and products at home and abroad. As people turn their attention to this direction, it is believed that government departments will issue a set of general AI ethics regulations to set an example for the whole industry in the near future.

Many people have questioned and discussed the relationship between artificial intelligence and humans. "*The Myth of the machine*" put forward a strong objection to the work of the machine, believing that the birth of the machine would make human beings lose their personality, thus making society mechanized. In recent years, the proposal and publicity of singularity theory have made people more worried about whether machines will completely replace humans. The core idea of this theory is that machines will soon surpass humans.

The author believes that the continuous progress of artificial intelligence is an indisputable fact. The feeling and movement of machines and the performance of computers will far surpass that of humans. This is the strength of machines, but it will not fundamentally have impact on human's posts and occupations. This is due to the following considerations. First, machines have their own advantages, while humans also have their own advantages, and this advantage is unmatched and imitated by machines in the short term. Human beings have the ability of thinking, can quickly extract and summarize laws from small data, and can make irrational decisions with limited resources. Human beings have intuition and can relate irrelevant things. Humans also have different internal processing methods from machines. Something that seems easy for humans to do may consume huge resources for machines. In 2012, Google trained the machine to spontaneously recognize cats from 10 million pictures. In 2016, the Google brain team trained the machine to automatically adjust the grip force according to the material of the object. This is a very simple task for a child, but it is quite hard to achieve it in the field of artificial intelligence. Perhaps, as explained by Moravec's paradox, high-level reasoning requires little calculation, but low-level sensorimotor skills require huge computing resources.

Secondly, humans and machines have not reached synchronous and symmetrical interaction at present, and there is still a time difference between the interaction. So far, it is still human beings who take the initiative and have irreversible advantages over machines. In his book "*The Essence of intelligence*", Piero Scaruffi once proposed that people establish rules and order in a chaotic nature because it is easier for human beings to survive and multiply in such an environment. The higher the structural degree of the environment is, the easier it is to manufacture the machines in it. On the contrary, the lower the structural degree of the environment is, the less likely it is to be replaced by machines. It can be seen that the emergence and development of machines are based on people's understanding and transformation of their environment. In turn, the development of machines has further promoted people's transformation and cognitive activities. This is just like the two ends of the balance. Simply removing either side will lead to an imbalance of the balance. Without human's guidance and transformation, the machine can only stay at the low-end level of mechanical and repetitive work. It will also make people constantly pursue a higher level of structure to make the machine move forward to a higher level when the machine works at a lower level. This is like an iterative rising process of the human-machine-human-machine cycle, in which human beings are always in a leading position. So machines can do human's work, but not replace humans.

Third, the rapid development of artificial intelligence has brought opportunities at the same time. It is true that the development of technology will bring some negative effects, but the advantages outweigh the disadvantages from the global perspective. The opportunities brought about by the development of new technologies are all around. The multiplicative effect illustrates that at least four jobs in other industries are correspondingly added for every job added in the high-tech field. Correspondingly, the ratio of traditional manufacturing is 1:1.4. We can see that with the rapid development of the artificial intelligence industry, related companies have sprung up rapidly, which has promoted the development of related industries (service industry and financial industry) as a whole, bringing more job opportunities meanwhile.

Moreover, all the technologies are not accomplished in a short time, but it is a gradual process. The earliest apes' tool producing, the later development of electric power, and the current internet age: the development and application of technology takes time to be ensured. Nowadays, some people in society are worried that the development of artificial intelligence will immediately rush to their own work, but they are actually a little "groundless". With history as a clear lesson, major technologic breakthroughs in history have not produced a devastating blow to human work. The birth of the steam engine replaced the traditional mule and horse, the birth of the printing press replaced the traditional scribe, the production of agricultural automation facilities replaced many farmers' jobs, but this did not result in the displacement of a large number of workers. On the contrary, people found what they originally belonged to. The job opportunities created by emerging technologies are higher than those that are replaced. Therefore, people do not have to worry too much about the problem of machines replacing human jobs.

In July 2016, a Tesla driverless car had a major accident that killed a driver at the scene. The accident quickly became the focus of the news media. People are not only concerned about the impact of this event itself, but also more worried about the machine as the subject of behavior execution and the mechanism for assuming responsibility after an accident. Who should be blamed for the accident, the machines that do the actual actions (without knowing what they are doing) or those who design or give orders, or both? If machines should be punished, how should they be dealt with? Should all memory of the machine be emptied like in West world, or should it be destroyed directly? At present, there is no relevant law to regulate and restrict it.

With the gradual popularization of smart products, people are becoming more and more dependent on them. In the human-machine hybrid environment interaction, people's tolerance for it has gradually increased.

Therefore, when there are some small errors in the system, people often attribute them to external factors, ignoring the accumulation of these small errors, and people always hope that they can be automatically repaired and returned to normal working conditions. Unfortunately, the black-box state of the machine does not present its own working state, resulting in a blank period of human cognition in human-computer interaction. When machines cannot repair themselves, the initiative is often handed over to humans, who are forced to participate in the cycle without knowing what happened or what to do about it. According to relevant surveys and studies, if people are under time and task pressure, they often have excessive cognitive load, which leads to avoidable mistakes. If it happens that a critical part goes wrong at this time, there is a great danger. Afterward, people tend to blame the people for their inaction and often ignore the responsibility of the machine side, which is biased. Perhaps as Perot said, 60%–80% of errors can be attributed to operator error. But when we look back at the errors again and again, we will find that operators are often confronted with unknown or even strange behaviors in system failures. Our past experience cannot help and we are just hindsight.

In fact, the author believes that there are three interaction modes in artificial intelligence, mode of people-in-the-loop, people-outside-the-loop, and the combination of the two. The mode of people-in-the-loop is called being in control, and people have greater initiative, so people have a sense of control over the entire system. The mode of people-outside-the-loop is called automation and people's initiative is entirely attributed to the machine. The third situation is that people can actively/passively enter the system. Most of the current so-called unmanned products have active mode/automatic mode switching. Among them, the passive mode is not desirable. As discussed before, the passive mode is unstable to the system in both time and space, and it is easy to cause unnecessary accidents.

There is also a special case that the accident is deliberately manipulated by the designer/manipulator. The most typical example is the military drone. To reduce its own casualties, the military tries to use drones instead of manned aircraft for military activities. The creation of drones is pulling the distance between the operator and responsibility farther and farther, and as drone missions become more complex, there are more and more manipulators behind the scenes, each of whom is only a small part of the "accident" So people's responsibilities are gradually diluted, and people became more comfortable with this kind of "killing" And many also believe that drones are smart enough to do as little harm to innocent civilians as possible compared with military personnel. Ironically, American drones have killed 2500 to 4000 people. About 1000 of them were civilians, and 200 were children. In 2012, Human Rights Watch highlighted in a report that fully autonomous weapons would increase harm to civilians and would be incompatible with humanitarianism. However, the current research on the ethics of military intelligent weapons is still at the theoretical level, and more efforts are needed to be practiced in actual military warfare.

To sum up, it can be seen that in some complex human-machine environment systems, the responsibility for accidents is difficult to define. Each person (or machine) is a part of the system and has completed a part of the function of the system, but the whole has produced irreparable errors. As for how humans and machines should coexist in artificial intelligence, the author will give some views in the following section. Through the above discussion and analysis, the author believes that artificial intelligence is far from having the concept of ethics (at least for now), which is sometimes only the corresponding concept of ethics for humans, and it is humans who impose the concept of ethics on machines. In the subconsciousness, people always regard the machine as a cooperative human, so they endow the machine with many words that did not belong to it, such as machine intelligence, machine ethics, machine emotion, and so on. In my opinion, these words are not wrong in themselves, because they reflect the high expectations people have on machines, expecting them to understand the minds of others like humans do and to interact naturally with humans. However, it is imperative to understand the parts of human ethics that can be structured, and only by this can we allow machines to learn and form their own ethical systems. Moreover, ethics is composed of ethics and morality, of which each part has its own meaning. The Chinese character "ethics" means human ethics, which is gradually formed by human beings in the long-term evolution and development and has a great cultural dependence. More importantly, ethics are situational. In other words, ethics are acceptable in one situation but difficult to understand in another situation. Therefore, how to solve the cross-scenario issue of ethics is also an issue that needs to be considered.

And it is worth mentioning that, as far as human-machine environment interaction is concerned, the machine refers not only to the machine and the simple computer, but it also includes the mechanism and the mechanization. The environment not only refers to the natural environment and social environment, but it also involves people's psychological environment. Simply focusing on one aspect will always cause fallacy of composition. The development of artificial intelligence technology is not only the development and progress of technology, but more importantly, it is because the mechanism and mechanization keep pace with the times. Because the development of the two complements each other, the rapid development of technology and the imperfect mechanism will restrict the development of technology. The current artificial intelligence ethics research has a bit of this meaning. The mechanization of human intelligence is still unclear, let alone the mechanism of machine intelligence. Moreover, at present, most of the machines pay attention to the external environment of people, namely, the natural environment and the social environment. The environmental data obtained by the machines from the sensors can comprehensively analyze the external environment of people, but it is difficult to have corresponding algorithms to analyze the human environment. In the internal psychological environment, people's psychological activities are intentional and motivated, which are also not possessed and understood by current machines. Therefore, for the development of artificial intelligence, the development of machines is not only the development of technology but also the continuous improvement of the mechanism. The development of machines that attempt to understand human implicit behavior is a further goal. Only when this goal is achieved can the human—computer environment interaction reach a higher level.

The ethical research of artificial intelligence is a product of the development of artificial intelligence technology to a certain extent. It includes not only the technology research of artificial intelligence, but also the exploration of the relationship between machines and humans, machines and the environments, and among humans, machines, and the environments. Consistent with many emerging disciplines, it has not got a long history but has developed rapidly. Especially in recent years, relying on the rise of deep learning and the emergence of some big events (AlphaGo victory over Shishi Lee), interest in AI itself as well as AI ethics research has risen sharply, and its related research and works have relatively increased. However, it can be expected that AI technology itself is far from the level of intelligence imagined, and the idea of spontaneously moving human ethics into the machine itself is extremely difficult to achieve. And if look back, we may find that AI always advances in ups and downs: it is a difficult problem to avoid how we guarantee that government funding remains at the same level as enthusiasm in peak or low cycles. These all require further research by current AI ethics experts.

In short, the ethical research of AI should not only consider the rapid development of machine technology but also consider the interactive subject-human thinking and cognition, so machines and humans can perform their own duties and promote each other. This is the prospect and trend of AI ethics research. Any technology is a double-edged sword, and AI technology is no exception. Al technology can help conceal the true and show the false in military affairs, build momentum and use momentum, and enhance the availability of products and systems in civil affairs. In recent years, however, it has been reported that a large number of artificial intelligence synthetic information occupied people's real life and virtual life (internet space), such as "nuisance phone" "praise irrigation" data pollution, synthetic sound, Al-generating video, images, things that do not exist, etc. In view of this, many experts believe that the rapid development of artificial intelligence makes real, virtual life space from real communication between people into intelligent, automated platform interaction and confrontation. So how to understand the phenomenon of artificial intelligence fraud and then how to change this phenomenon become important questions on which people to think deeply.

Machines are essentially man-made objects and cannot be faked. They are more similar to the function that turns the input into the output, only caring about how much it counts but not asking whether it is right or wrong. More importantly, besides calculus, the variation of data and information is the evolution and deduction in which people participate. So artificial intelligence fraud is human fraud in nature, which is a fraud that people conduct through compiled programs and equipment. Like much magic, it is just a new algorithm that is combining some old formulas, theorems with a new situation (such as a well-trained GAN [generation confrontation network technology] with rich data resources; it is not difficult to achieve realistic photos, video, and text materials). Its purpose is to make people have a sense of confusion, perceptual deception, to achieve a situation in which the truth is submerged and the false succeeds.

However, the world is made up of facts (relationships), not things (attributes). Fundamentally, even if the most basic units that make up things are found, the machine cannot really understand the interaction between the units. And these interactions are the biggest secret of the world. For example, modern physicists have found that the human body is the same as the basic physical particle structure of water, stone, and other things, but it still cannot explain why humans can produce consciousness and emotion. If we grasp the essence that "the connection between things is the source of the world" most cases of AI fraud can be seen through and beaten.

Artificial intelligence fraud technology can filter, screen, and exclude deep situation awareness or context perception technology accordingly. For example, just as a normal person will not easily do abnormal things, a normal institution will not be recklessly lawless. However, there may be accidents in extraordinary periods and extraordinary situations, so it is not difficult to identify general cases. But it is difficult to conduct subtle distinction and screening under special situations. In addition to the conventional anticounterfeiting techniques, we also need to develop new technologies and tools of deep situation awareness, identify and intervene in the early stage of counterfeiting as much as possible, and carry out in-depth analysis and response from the stages of state, trend, sense, and percept, such as developing watermarks hidden in video and generating antagonistic neural networks of hidden information. It has deep situation awareness of voice, video, image, telephone, network analyzer, antipixel attack, and other technologies. In addition, we can also study the corresponding management emergency mechanism methods and means, strengthen the relevant legal and moral management, formulate relevant laws and regulations as soon as possible, and popularize the corresponding knowledge, so the relevant applications of antiartificial intelligence counterfeiting technology can be truly applied to relevant units and thousands of households, truly realizing the antiartificial intelligence counterfeiting ecologic chain of human-machine-environment system linkage.

The fake cannot be true, and the real cannot be fake. The devil is a foot high, and the Tao is a foot high. After all, the best artificial intelligence is artificial, and the fake made by artificial intelligence should not be complete. Human beings themselves are the best hunters to deal with incompleteness. What bad people cannot get in the real world, they will not get in the virtual world either; after all, a bad-person-centered situation is against the most fundamental interests of most people.

CHAPTER 10

Rethinking intelligence of human-machine hybrid

Contents

1. The origin and future development direction of artificial intelligence

Artificial intelligence (AI) really originated in Europe and was first expressed in philosophical and mathematical forms, such as "who am I?" in ancient Greek philosophy and "universal words + rational calculus" in Leibnizian mathematics, etc. The concept of AI was introduced at the Dartmouth College Summer Forum in 1956 based on the idea of a British mathematician. Over the next 60 years, artificial intelligence has made remarkable progress with the development of machine learning, data mining, deep learning, as well as other technologies. Three main theoretical ideas have emerged during this period, namely connectionism represented by neural networks, behaviorism represented by enhanced learning, and symbolism represented by knowledge atlas (expert system). Recently,

Integrated Human-Machine Intelligence
ISBN 978-0-323-99562-7
https://doi.org/10.1016/B978-0-323-99562-7.00010-3

DARPA, the advanced technology agency of the US Department of Defense, analyzed the development stage of AI technology based on its technology characteristics, and it believes that AI has gone through the first and second waves and will usher in a third wave. The first wave of AI technology started in the early 1960s, characterized by "manual knowledge" which represented knowledge in a particular domain by creating a set of logical rules, reasoning about tightly defined problems, and having no learning capability and a weak ability to handle uncertainty. The second wave of AI technology began in the late 1960s and was characterized by "statistical learning" building statistical models for specific problem domains and training them with big data, with a low degree of inference but no contextual capability. The third wave of AI technology is characterized by "adaptive environment" (contextual adaptation), which is continuously learnable and interpretable, building systems that can generate explanatory models for real-world phenomena, where machines can communicate naturally with humans, and where systems can learn and reason when they encounter new tasks and situations. The ability to learn will be the core driver of the third wave of AI technology. Based on this, we think and analyze that the fourth wave of AI technology will be characterized by "active adaptation to the environment" (wider contextual adaptation), with sustainable + unsustainable learning and explainable + unexplainable model systems that can generate active and moderate explanations for real and virtual world phenomena. The system is capable of mutual learning and reasoning when new tasks and situations are encountered. Active mutual learning, mutual understanding, and mutual assistance in human-machine hybrid and mutual fusion capability will be the core drivers of the fourth wave of AI technology.

Objectively speaking, artificial intelligence is only a describable and programmable part of human intelligence, while human intelligence is the product of the interaction between humans, machines (object), and environmental systems. The mechanism of intelligence generation, in short, is the result of the superposition of the interaction of a human-machine-environment system (machine is a man-made object), and it is composed of the state, situation, and potential (referred to trend) that are derived from the superposition of the changing states of human, machine, and various environments. The state of the three changes is different, either good or bad, high or low, prosperous or inverse, and the generation of embodied intelligence is determined by the degree of harmony and resonance of the system state and trend. There is a constructive and destructive interference effect among these three sides, which can be enhanced or eliminated. The intelligence will be strong when the three coincide and will be weak otherwise. How to tune the common frequency of the three is the key to human-machine hybrid intelligence. The next development may be to realize machine cognition and eventually machine awakening by means of human-machine hybrid intelligence. Contemporary artificial intelligence has developed from the initial completely artificially compiled machine automation to artificially precompiled machine learning, and the next development may be to achieve machine cognition and eventually machine awakening through the method of human-machine hybrid intelligence.

2. The key issues in the future of human-machine hybrid intelligence

At present, the development of human-machine hybrid intelligence is still in the primary stage. The first key problem of human-machine hybrid intelligence, which is also the most important problem, lies in how to combine the computing ability of the machine with the cognitive ability of the human. The current human-machine hybrid in the application stage has a clear division of labor between human and machine, which does not produce an effective combined effect. Humans continue to expand their cognitive abilities in acquired learning, so they are able to understand the development of the situation more precisely in a complex environment. Through the ability of association, humans can have the ability of cross-domain combination, and this cognitive association ability is precisely what is missing. How to make machines produce this ability is a breakthrough to achieve true intelligence. Integrated information theory from Giulio Tononi suggests that a conscious system must be information integrated at a high speed. At the same time, the evolution of machines with cognitive-like capabilities requires the existence of a common consciousness between humans and machines. Therefore, a high-speed and effective two-way information interaction between humans and machines must be established. The basic of cognition lies in abstraction, and for machines, the ability to abstract determines the limiting environment of the problem. The more abstract the thinking representation is, the more it can adapt to different situations. In the meantime, high abstraction ability will also bring a more general migration ability, thereby breaking through the limitations of thinking. John McCarthy, 1971 Turing Award winner, put it this way, "As with all specialized theories, all science is embodied in experimentation.

When you try to prove these theories, you go back to attempt reasoning, because common sense guides your experiments" Common sense is the unstructured multimodal information, the complex of support, and in cognition is the prior knowledge of humans, which is precisely what computers ignore with their information input. Therefore, studying knowledge itself, knowledge type, and knowledge principle is also the key to break through the combination of cognition and calculation.

Another key issue of human-machine hybrid is the mixed axiomatic and nonaxiomatic reasoning, where intuition and reason are combined in decision-making. Axioms are the theoretical foundation in the history of mathematics, and logical deduction is the most core method in the process of scientific research. Similarly, the computer still runs in a rigorous algorithmic language. But human decision-making is different from this process in that human associative ability also relies on analogical reasoning, which is part of nonaxiomatic reasoning. Nonaxiomatic reasoning determines strong perceptual problems in weak situational situations. This learning method relies on prior knowledge and is implemented by using big data with probability. And implementing nonaxiomatic reasoning for machines is one of the differences between humans and machines. More so, it is an important way for human emotions to be realized on machines. Through prior knowledge humans generate intuition, and rational analysis is the opposite of intuition. Machines are always processing data rationally, and how to make machines produce intuitive capabilities is the key to a smooth human-machine hybrid. The combination of axiomatic and nonaxiomatic reasoning, intuition, and rational decision-making will be an important research direction to address the intelligent output of the human-machine hybrid.

The key issues of human-machine hybrid intelligence also include intervention issues, which reflect the problem of timing and manner of a human-machine hybrid. This problem arises especially when there is an asymmetry between human and machine in terms of perceived information, and when there is a conflict between human and machine in terms of the direction of decision-making. Meanwhile, the intervention problem in a human-machine hybrid is also reflected in team situation awareness, and the weight of team tasks is also gradually biased toward team situation awareness of human-machine groups. Interactions between team members in team situation awareness include acceptance, tolerance, trust, matching, scheduling, switching, and persuasion, which are the conditions that make the cooperation between team members produce the effect of team over individual. The intervention problem of a human-machine hybrid has the same complexity as the human to human interaction problem. From the technical perspective of view, the human-machine hybrid intelligence is by no means only a mathematical simulation modeling problem, but also a psychologic ergonomics problem, and it should also be a problem of experimental statistical experience fitting.

The final key issue of human-machine hybrid intelligence is ethics. The origin of human values is ethics. It is easy to see from the inconsistency of different individuals in team situation awareness that humans themselves possess many ethical and moral dilemmas, and in addition, the emergence of artificial intelligence has also brought human thinking about the ethical issues of artificial intelligence. Meanwhile, the category of human-machine hybrid intelligence is the key to the ethical problem of it. The ethics of human-machine hybrid intelligence includes not only the ethics of artificial intelligence, which includes the impact of AI ideas on practical legal problems, but also the definition of a human-machine hybrid and the idea of whether the resulting behavior is attributable to human or machine thinking. In addition to thinking, the need to face specific legal responsibilities for actions generated by devices as part of humans in human-machine hybrid intelligence is also one of the most important issues in the next development of human-machine hybrid intelligence.

3. The future development direction of human-machine hybrid intelligence

3.1 Information fusion and human-machine hybrid intelligence

Information fusion originates from data fusion, or data fusion is the first stage of information fusion. Data fusion uses the data and results from multisensor detection to form more accurate and credible conclusions and quality that cannot be obtained from a single sensor. The earliest data fusion was limited to the differences in hardware devices that required the addition of manual grooming, but despite this, the sensors still had problems with timeliness and accuracy due to hardware problems, which had a successive impact on subsequent work, which led to a gradual shift in research toward a fusion approach. The second phase of information fusion development incorporates other information sources in addition to the use of multisensor detection data. Moreover, compared with the fusion of sensor data, the methods and techniques of information fusion from multiple information sources are more complicated. There is a need to move from statistics and structured models to unstructured models, as well as artificial intelligence techniques and knowledge-based systems. Additionally, information fusion is constantly entering into advanced perception fields such as situation and impact estimation. Information fusion models at this stage still use only massive data scale, fast and dynamic data systems, multidata types, and low data value density.

Information fusion is a key link of human-machine hybrid intelligence. In the current two stages, information fusion, both in theory and in technology and application implementation, lies only in the attempt to build a product that runs automatically, embedded in an application system or directly applied to the corresponding business activities as a system. In contrast, advanced fusion problems such as target identification, situation estimation, and impact estimation, which cannot be solved by the traditional structured mathematical models and methods such as statistics, calculation methods, mathematical programming, and various information processing algorithms, resort to uncertainty processing and artificial intelligence techniques, while the current development of uncertainty processing techniques, especially AI techniques, is far from advanced information (e.g., human needs). The shift from "being" to "should" problems in dealing with uncertainty is where people excel. Adding human choice, judgment, and action management in the operation process of information fusion system is the key to enable the information fusion intelligence to achieve qualitative changes in the field of advanced perception in observation, judgment, analysis, and decision-making in information fusion intelligence.

3.2 Situation awareness and human-machine hybrid intelligence

The concept of situation awareness first appeared in aviation psychology to describe a pilot's understanding of the situations in an operational flight mission. The classical theory of situation awareness is the three-level model proposed by Endsley in 1995, which is defined as human ability to perceive (perception), comprehensively understand (comprehension), and predict (projection) various elements of the environment in a given space and time. Over the past 2 decades, the research on situation awareness has gradually expanded to the fields of civil aviation pilots, air traffic controllers, nuclear power plant operators, military commanders, etc. In these areas, the situation awareness of operators is a key factor in decision quality and operational performance. And having good situation awareness plays a critical role in decision-making in complex and dynamic systems such as aviation, air traffic control, and aircraft piloting.

The concept of situation awareness emerges in the work of human-machine collaboration. In the three-level model of situation awareness, perception is the acquisition of information, which mainly relies on the sensors of machine under high-load cognitive conditions, and later it is processed by the computer and presented to the operator. The machine plays an important role in the perception phase of the three-level model. In the decision-making stage after prediction, collaborative judgment and analysis between machine and human are also required. The separation between human and machine in the three stages is a flaw in this model. To promote the integration of human and machine in situation awareness is the key to achieving good performance in situation perception. The components of a specific situation composed of human, machine, and environment often change rapidly, so sufficient time and enough information are needed in this fast-paced situation evolution to form a comprehensive perception and understanding of the situation. Likewise, in the case of insufficient state and trend, human-machine hybrid intelligence provides a solution for strong perception by virtue of prior knowledge through big data processing and analysis in assisting operators' decision-making.

3.3 Autonomy and human-machine hybrid intelligence

Automation has been applied in various systems, and it usually involves needs for software to provide logical steps and operations. Automation is traditionally defined as follows: "The system operates without or with little human operator involvement. However, system performance is only limited to specific operations that are [sic] designed to be performed." Autonomy, compared with automated systems, involves the use of additional sensors and more sophisticated software to provide a higher level of automated behavior over a wider range of operating conditions and environmental factors, as well as a wider range of functions or activities. An autonomous system has a certain degree of autonomous behavior (employing human decision-making). Software methods can be extended to computing logic-based (or more generally rule-based) methods to include computing intelligence (e.g., fuzzy logic, neural networks, Bayesian networks). Alternatively, learning algorithms can provide the ability to learn and adapt to changing environments. Autonomy is a major extension of automation where high-level task-oriented commands will be successfully executed in a variety of situations that may not be fully expected, just as we currently expect intelligent people to run the same execution permissions when given sufficient independence and tasks. Autonomy is well designed and highly automated.

However, autonomous systems are always faced with several common problems. The problem of the design ability of autonomous systems is the balance of autonomy between human and automation. When facing new and immutable environments, there are games between lightly repetitive work and reliably repetitive work, discontinuity and consistency, unpredictability and predictability; the operators always do not understand what automation is doing due to their situation awareness of autonomous systems and advanced automation, so it is needed to provide pilots with appropriate levels of participation to keep them not disconnected from the autonomous system. The problem of an auxiliary system is that automated auxiliary systems often give operators a high sense of trust. Just like asking experts for help, the label of expert itself brings a kind of trust. But the truth is, the real evaluation of results should rely on the answer to the problem itself, not the external label. For the operator, the same auxiliary system will bring the same trust, but this kind of trust will bring disaster in the case of deviation. The trust problem is that it is affected by a variety of factors from the system, the people, and the situation. Wrong judgments brought about by the situation will quickly reduce an operator's trust for the system, and it is very important to make the operator trust the autonomous system and further make better operations in this psychologic environment.

How to solve the problem of balance between human and automation is an important issue in human-machine hybrid intelligence, as well as the issue of trust between human and machine. Humans should be centered under autonomous systems, which do not need to seek to completely replace people with machines, in which human control and command are indispensable. Therefore, more flexible autonomy and autonomy switching are needed. As the system's abilities improve, so does the level of autonomy. Decision-making assistance provides operators with potential options, and the operator can make appropriate interventions during supervision and control. The level of autonomy system used in a specific situation changes dynamically. For example, a high level of automation can be used in a low-risk situation, while human involvement in a self-service system should be regulated after a change in risk. Shared human and machine situation awareness is also important. Even people in the same environment with the same display will have different goals and mental models, and thus different predictions about the future. Autonomous systems get information through sensors to understand the world in a different way than people do.

Therefore, there is a need to share the situation awareness of both humans and machines. This is reflected in goal congruence, function allocation, and reallocation, finding a balance between human and machine tasks, and decision-making communication, including strategy, plans, and actions, and task alignment, as tasks often require close dependence on both parties, so from these four aspects, there is a need to keep the autonomous system and human situation awareness aligned.

4. Difficulties facing human-machine hybrid intelligence

The concept of "intelligence" implies the relationship of the individual, the finite to the whole, and the infinite. In response to the advent of the age of intelligence, it has been suggested that "the spatio-temporal principles of behavior that have existed since ancient times need to be considered and understood from a completely different perspective," such as the traditional relationship between people, objects, and the environment. When people carry out an intelligent activity, they generally make corrections or adjustments at key points or critical points according to changes in the external environment and assign weights to various data and information knowledge processing in real time through local and global short-, medium-, and long-term optimization expectations, which is more of a programmed + nonprogrammed hybrid process. Machine intelligence, on the other hand, is difficult to achieve such a randomized hybrid response, and the deterministic programmatic mark is more prominent, and the better intelligent systems like AlphaGo/yuan/star mainly win in the speed and accuracy of computation with clear boundaries. True intelligence is not only adaptive, but more importantly, nonadaptive, and thus creates a new possibility. The disadvantage of Turing machines is that there is only stimulus-response and no choice, and only compliance and no assimilation mechanisms.

Is the world made up of connections or attributes? This is a question worth thinking about. It should be composed of both. It is said in Chapter 40 of the *Tao Te Ching*, "The opposite is the movement of the Tao; the weak is the use of the Tao. Everything under heaven is born from existence, and existence is born from nothingness" This sentence is the concentration of this idea. Here, it is said that "the opposite" is relative to "the positive" which also means "round-trip" and "the weak" is relative to "the strong" and only with the opposite can there be the positive, which is called yin and yang. Here we use a very elaborate phrase to explain the Tao, which is "one yin and one yang is the Tao" The weak and the strong are yin and yang. Only with yin and yang can the Tao move, and only then can there be interaction.

The essence of informatization is calculating fact, while intelligence is cognitive value. From data to information to knowledge (structure) is cognitive computing, and from knowledge to information to data (deconstruction) is computing cognition. If intelligence is considered language, then AI is like grammar and human intelligence is more like semantics and pragmatics. Grammar is based on rules, statistics, and probability, while semantics and pragmatics are based on a kind of convention between people using meaningful elements, and subconscious conventions that are more transboundary and flexible than grammar, and people have not yet formed effective rule cognition about its laws, so it becomes complex. Symbolization is the representation of normative grammar, and contextualization is the basis of naturalistic semantics. Individual and group environments have reductive components as well as novel elements, and one of the difficulties in understanding intelligence is the intertwined interference and influence of the coexistence of one inside and many outside. The transformation of sending anyone any time, place, and information to conveying them to the right person at the right time and place is one of the manifestations of intelligence. On the whole, the human is the ascending dimension of the machine, and the machine is the descending dimension of the human; in the local, it is the other way around. This is because the whole involves heterogeneous things, nonfamily similarities, while the local is the opposite. For human intelligent systems, the role of Go chess is still only a partial part.

The most underlying technology of artificial intelligence is the "0, 1" binary logic of diodes, and the most underlying technology of human intelligence is the multiplicity of human intentions (nonlogic). Human intelligence, on the other hand, is art, and artificial intelligence is primarily technology. Artificial intelligence is just a tool, but many people treat it as the omnipotent key, even some imagine it as the omnipotent Sun Wukong and Santa Claus, while ignoring the role of human intelligence. Human intelligence is a kind of intelligence that involves more sensuality (especially courage); in the case of rapid changes in emergency situations, a person is governed by emotions rather than thinking, and thus sanity needs to evoke the quality of courage, which will maintain necessary sanity in the last action. In human intelligence, we can often see the creative tension between order and disorder, such as in many situations, where the same thing you see (e.g., an apple or 1 h) will be different. Also, if seeing actively, passively, and semiactively, the results will also be different. Artificial intelligence often tends to form bias, extracting prior and common sense from the knowledge graph of rules and introducing it into the generative model as a constraint, which may make the intelligent program run much worse, so how to combine the fuzzy perception and recognition of human with the precise perception and recognition of machine will be worthwhile to think about.

4.1 The problem of cognitive inconsistency between human and machine

The main reason why human-machine intelligence is difficult to integrate lies in the inconsistency of space, time, and cognition. The information and knowledge processed by humans are capable of variation: a thing or fact they represent is not only itself but also other things, facts, which have always been relative, while the data identification processed by machines lack such relative variability. More importantly, human perception of time and space is intentional and subjective expectation (should), while machine perception of time and space is biased toward formalization and objective existence (being). The two are not in the same dimension, so there is a strong inconsistency. Human cognition is focused on the psychologic dimension and is subjective, while machine cognition is biased toward the physical dimension and is objective. In terms of cognition, human learning, reasoning, and judgment are random and adaptable, and the law changes with time and events, while the learning, reasoning, and judgment mechanisms of machines are formulated or selected by specific designers for specific spatiotemporal tasks, and they are often not fully consistent with the user's intention in the current spatiotemporal task and less variable. This inconsistency includes both the inconsistency between subjective human expectations and objective data feedback from the machine and the inconsistency between subjective human expectations and objective facts.

Many things appear to be nonlogical problems, such as the many cases of the weak prevailing over the strong, but in fact, they are actually logical problems. The weakness in these cases is relative, but in the local the strong always wins over the weak, so there are many logical relationships in nonlogic. Similarly, many logical problems also have nonlogical parts, such as some cases that are logical apparently but unreasonable; in fact, these reasons are variable and incomplete, which are bound by preconditions and boundary conditions. When these many preconditions and boundary conditions occur with some minor changes, they will naturally be out of reason. It can be seen that logic and nonlogic coexist in things and is the root of order and disorder, in which the interaction and organization are the focus of research on human-machine hybrid intelligence but also the difficulty of human-machine hybrid intelligence.

Another key issue of human-machine hybrid is the mixed axiomatic and nonaxiomatic reasoning, where intuition and reason are combined in decision-making. Axioms are the theoretical foundation in the history of mathematics, and logical deduction is the most core method in the process of scientific research. Similarly, the computer still runs in a rigorous algorithmic language. But human decision-making is different from this process in that human associative ability also relies on analogical reasoning, which is part of nonaxiomatic reasoning. Nonaxiomatic reasoning determines strong perceptual problems in weak situational situations. This learning method relies on prior knowledge and is implemented by using big data with probability. And implementing nonaxiomatic reasoning for machines is one of the differences between humans and machines. More so, it is an important way for human emotions to be realized in machines. Through prior knowledge humans generate intuition, and rational analysis is the opposite of intuition. Machines are always processing data rationally, and how to make machines produce intuitive capabilities is the key to a smooth human-machine hybrid. The combination of axiomatic and nonaxiomatic reasoning, intuition, and rational decision-making will be an important research direction to address the intelligent output of a human-machine hybrid.

4.2 Intentionality and formalization issues

British computer scientist and AI philosopher Margaret Borden proposed early on that the core and bottleneck of AI lies in the organic combination of intentionality and formalization, and there is still no breakthrough to this day, which is actually the difficulty of human-machine hybrid intelligence. Among the human-machine hybrid products currently put into application, the division of labor between human and machine is clear but not organically combined. Humans are able to make better predictions about the development of situation in the case of incomplete environmental information and resources, and this is because humans can continuously enhance their cognitive ability in the acquired learning. Machines do not have the ability of association, while humans can generate the ability of cross-domain integration precisely through association. Therefore, the key to realizing true intelligence is how to make machines have the ability to associate.

Intentionality is the description of intrinsic perception (mental process, purpose, expectation), and formalization is the description of extrinsic perception (physical mechanism, feedback). Human-machine hybrid intelligence and deep situation awareness are a combination of intentionality and formalization. Formalization tends more to allow people to have an intuitive spatial cognition of things, and extending that spatial perception to the description of time is intentionality. Formalization is the state, so intentionality is the trend. Human-machine hybrid is about forming a holistic description of internal and external, subjective and objective, cognitive and behavioral perceptions, and forming a model that can describe human mental processes, purposes, and expectations as well as the physical mechanism and feedback of the machine.

The current difficulty in the field of intelligence is the degree to which human intentionality differs from behavior, which can be objectively formalized, while intentionality is subjectively implicit. For an intelligent system to form and exist, its internal components must possess the ability of both attracting and repelling each other in their nature or operation laws, to converge and evade, combine and separate, integrate and disconnect. Intentionality in human-machine hybrid intelligence is the bridge between facts and values, and formalization can realize this intentionality to a certain extent.

4.3 The ethical issues of Hume's question

The final key issue in human-machine hybrid intelligence is ethics. The origin of human values is ethics. Humans themselves possess many ethical and moral dilemmas, and the emergence of artificial intelligence has brought humans to think about the ethical issues of artificial intelligence. At the same time, one of the key ethical issues of human-machine hybrid intelligence is the category. The ethical issues of human-machine hybrid intelligence include the ethics of artificial intelligence and the responsibility of the human-machine hybrid, which is an important issue for the future development of human-machine hybrid intelligence.

Hume's question proposed that value cannot be deduced from facts, but the world is a mixed world of facts and values, so can facts be deduced from values? Chinese characters are the concentrated manifestation of intelligence, tangible and intentional, such as the sun (日), the moon (月), and person (人). Western characters are often intangible and unintentional, logical analogies. The essence of intelligence is to unite intentionality with formalization, so the process of Chinese characters from pictogram to metaphor is a brief history of the development of natural human intelligence. The radicals of Chinese characters are a kind of encapsulation, bringing together strongly related words. If human character creation is the accumulation of encapsulated linguistic representations, then human intelligence creation is the topological extension of thought and consciousness. Intelligence is not an encyclopedia, but it contains a lot of fiction and imagination, not only classification, but also merging categories, not only merging similar items, but also merging heterogeneous items, so the top-level design of intelligent product systems is very important. Artificial intelligence is generally a logical (family similarity) relationship, and human intelligence is often nonlogical (nonfamily similarity). The future intelligence is the fusion of human intelligence and machine intelligence in a specific environment, that is, human-machine hybrid intelligence. Human-machine hybrid intelligence is not artificial intelligence, much less machine learning algorithms. Similarly, artificial intelligence and machine learning algorithms are not human-machine hybrid intelligence, which is the mutual fusion of human-machine environment, which is the "zhiji" (seeing the first signs), "qushi" (seizing the opportunity), and "biantong" (adapt to circumstances) in the *Book of Changes*. Human-machine hybrid intelligence is a follow-through, not a given, in which the "percept" in "knowing yourself and your enemy" is not simply the situational "perception" but also the situational "cognition" which is the process from trend to state, and perception is the process from state to trend. Cognition focuses on recognition, the flow process of information input, processing, and output; perception focuses on sense, the input filtering process of data information, and cognition involves past perceptions such as a priori and experience, so situation awareness also includes the previous one. Artificial intelligence is a double-edged sword: that is, the more fine and accurate the calculation is, the greater is the danger because the bad guys can hide the truth and deceive, so the organic integration of human and machine intelligence is more important. Objectively speaking, current AI is basically automation + statistical probability. Simply put, the drawback of inductive deduction is that completeness is explained by incompleteness.

Picasso once revealed, "Painting is not an aesthetic process, but a kind of magic, a way of gaining power, which overrides our fears and desires" If you understand Picasso's work, you will be able to understand the "magic" that Picasso wanted to express and apply it to other areas of life, especially in the field of intelligence and human-machine hybrid intelligence.

It is important to note that the Hume problem has not really been solved yet. Because "value" is relative and varies from person to person, this problem can never be truly solved, as has been discussed earlier. Although materialists want to carry materialism into the spiritual realm, this can never be done. For spirit and matter are, in essence, completely different things, one subjective and the other objective. The "brain-in-a-barrel experiment" (proposed by the British philosopher Putnam, also translated as "brain-in-a-jar" in some versions) often used by skeptics to describe human's knowledge of the world is in fact only a subjective judgment. Whether this judgment is consistent with the real "objective world" or not, one can never know. Although some materialists like to use "countless practice" to prove that subjectivity and objectivity can eventually achieve this consistency, in reality, "countless practice" is not possible. That is why it is just a fantasy.

5. Difficulties of human-machine hybrid intelligence: research on deep situation awareness

The definition of situation awareness will not be elaborated here. The term "situation awareness" was first introduced in World War I and has since been widely used in psychology as "situation awareness" Endsley's redefinition of situation awareness in 1988 and his famous three-level model of situation awareness in 1995 marked the migration of situation awareness to engineering. In 2003, Wickens proposed the attention-situation awareness model (A-SA model), and in 2010, Hooey introduced the situation element into the study of situation awareness, which marks the transition from subjective data-driven to objective data-driven, qualitative to quantitative analysis in situation awareness research. In recent years, with the rapid development of artificial intelligence-related technologies, cyber situation awareness (CSA) has become a hot research topic in the field of cybersecurity. CSA seems to be a research method, rather than a methodology that can guide people to understand and transform the world. There are still many shortcomings in current situation awareness theories and techniques, mainly because of failure to combine the process of human psychologic activity with the external manifestations of machines and the situational elements in the environment. In view of this, this section attempts to introduce the concept of deep situation awareness, which is described subsequently.

Deep situation awareness means "situation awareness is a kind of human-machine intelligence, where both human and machine intelligence (artificial intelligence) are included" It is signifier + signified and involves both the properties of things (signifier, feeling) and the relationship between them (signified, perception). It is able to understand both the meaning and the voice beyond the words. It is based on Endsley's situation awareness (including information input, processing, and output), and it analyzes the overall system trend including human, machine (object), environment (nature, society), and their interrelationships, with both "soft/hard" regulatory feedback mechanisms, including self-organization, self-adaptation and other-organization, and mutual adaptation, including both local quantitative calculation and prediction, and overall qualitative calculation and evaluation, and it is a kind of expectation-selection-prediction-control system with autonomous and automatic information correction and compensation of diffusion effect.

In Wiener's book "*Cybernetics: Or Control and Communication in the Animal and the Machine*" Wiener sees cybernetics as the science of studying the laws of machines, life, and society, and the science of how individuals (which may be living beings or machines) maintain homeostasis in a dynamic environment. The ideas and methods of cybernetics have been deeply influential on the research in the fields of social science and natural science. Wiener suggested in the book that "the core of control is feedback, and feedback is the purposive behavior of people." However, in revealing the natural existence of machines, cybernetics not only completely shuts out the social mega-machine, of which it itself is but a time period and a component, but it also completely shields the crucial issue of organizational generativity, which is inherent in all physical, biologic, and social machines except man-made ones.

In fact, replacing the concept of "purposive behavior" which is specific to living beings, with the concept of "feedback" and considering the working behavior of machines designed according to the feedback principle as a purposive behavior, does not break through the conceptual barrier between living bodies (humans) and nonliving bodies (machines). The reason is simple: human "purposive behavior" is divided into simple explicit and complex recessive, and the former can be approximately equivalent to the "feedback" of nonliving machines (stimulus-response), but the later intentionality is far from being able to be replaced by the "feedback" approximation, because this intentionality can be delayed, increased, decreased, and concentrated, and it is more accurate to be defined as "reflection" However, the concept of "reflection" is difficult to be endowed by a nonliving machine (stimulus-selection-response). The purposefulness of "reflection" can be characterized by subjective valence, which will be another key aspect of the human-machine hybrid. Value will be composed of both attractors and motives. Reflection is an unproductive feedback or an organized feedback. Autonomy is organized adaptation or organized adaptation. Accordingly, we combine Endsley's three-level model of situation awareness with Wiener's idea of "feedback" and propose a model of deep situation awareness based on "feedback" as shown in Fig. 10.1.

Theoretical models of deep situation awareness process information differently in different contexts, and previous studies on situation awareness have well illustrated the real-time nature of situation awareness; that is, situation awareness is constantly updated and iterated over time. Therefore, we try to refine the situation awareness and propose a deep situation awareness theoretical framework based on recurrent neural networks (RNNs), as shown in Fig. 10.2.

We define "state" in situation awareness as the subjective and objective data that characterize the state of an individual in the human-machine-environment system. "Trend" is defined as the development trend of events; the awareness of "state" is sense; and "percept" is defined as the

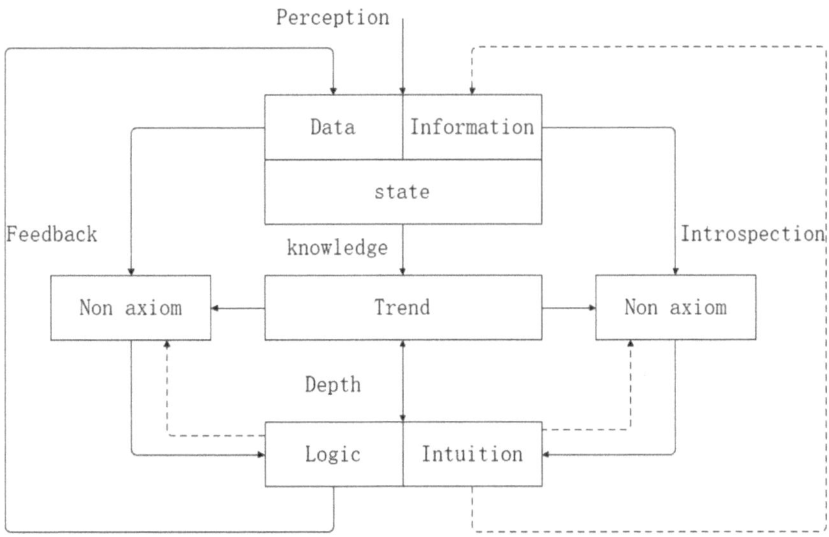

Figure 10.1 Deep situation awareness model. *(Based on "feedback.")*

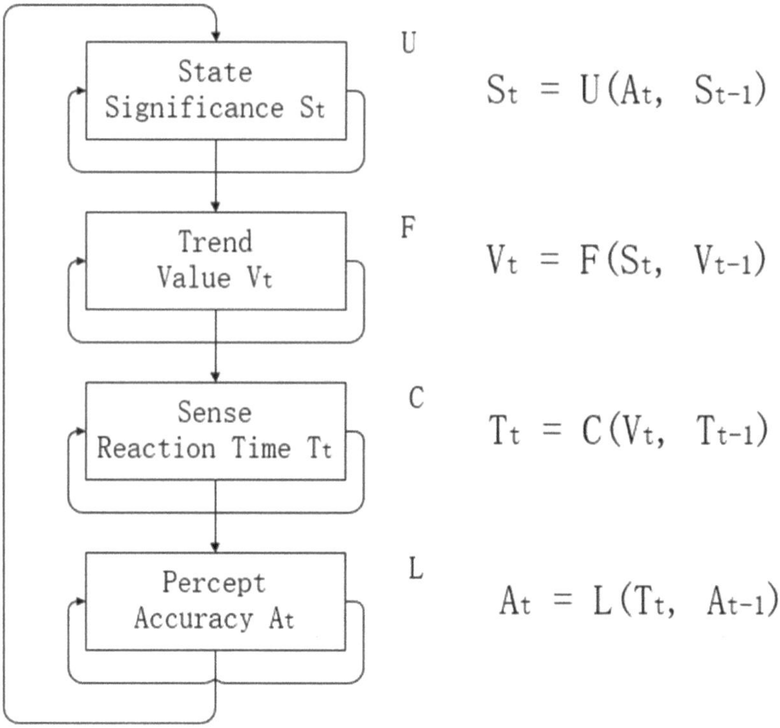

Figure 10.2 Theoretical framework of deep situation awareness. *(Based on RNNs.)*

understanding of "trend" This theoretical framework is designed to assist people to better "sense the state" and "know the trend" To obtain data, it is necessary to introduce objective data, and according to previous research, we can formalize state as salience, trend as value, sense as response time, and percept as accuracy. The sense state focuses on timeliness, while the percept trend tends to be more effective.

"I think, therefore I am" is the starting point of Descartes' dualistic epistemology, and its end as well; that is, the only certainty is the experience of "I." According to cognitive science, optimization is not possible because in most cases the cognitive capacity of a person is limited. The participant must also understand his goal equation, which requires another large cognitive prerequisite, and as the participant discovers their environment, systematically describing this goal equation becomes extremely complex. Knowing oneself and knowing others are inseparable, and one cannot know oneself without knowing others; nothing can explain itself by itself, but only from other references can one perceive, understand, discover, describe, and define oneself (who I am, where I come from, and where I am going), so one can argue that the self does not exist, and one cannot explain himself if there is no environment or reference, just as the concept of "I" cannot be defined as "I am what I am" Further, self-awareness may not exist, but it is a product of interaction, which can cross the logical relations of time and space. In fact, all autonomous systems are involuntary, but only to different degrees of visibility. Later, Descartes formalized his philosophical view into the famous binary right-angle co-ordinate system.

According to Descartes' view, the deep situation perception virtual-real reference system can be divided into different human-machine hybrid state (thing) reference systems, trend (fact) reference systems, sense (significance) reference systems, and percept (value) reference systems, and the correct awareness and decision-making behavior are possible only when these virtual-real reference systems are mostly consistent or not essentially contradictory.

Only when distinguishing one thing from others can one be recognized. Only by distinguishing a person's knowledge or belief state from that of others can man be understood. One of the most difficult and important philosophical tasks is to specify two types of characteristics of the world, namely, those intrinsic characteristics that exist independently of any observer and those extrinsic characteristics that exist relative to the observer or user. For example, an object has mass (for whomever) and the object is a bathtub (it can also be a water tank, a decorative tank, or a grain tank). The next step in the study of deep situation awareness systems is to apply them specifically to one or some situations, so we can test their effectiveness and reliability.

Many of you may have seen "*The Matrix*" a science fiction movie about artificial intelligence. Indeed, the current technologies related to artificial intelligence (big data, machine learning, deep learning, etc.) are based on mathematical theories such as matrix theory and probability theory, and they have facilitated people's production and life and have even led to social changes to some extent. The achievements of artificial intelligence have been made possible by 400 years of unremitting pursuit of mathematics since the 17th century, but what has been neglected in today's artificial intelligence, and what may help people break through the bottleneck of today, is the thousands of years research on people's cognition of the world and the study of self-reflection. Therefore, how to organically combine natural science and social science is the research focus of the next generation of artificial intelligence technology.

Man-made machines exist but have no ego. The self is born from the constant interaction, organization, and generation of one's own existence. If there is no active interaction or organization, there is no self, and without self, there can be no sense of self others. Knowing self, knowing others, and sensibility cannot be linked to rationality, no objectivity can form subjectivity, and no facts can derive value. Although intelligence is a complex system problem with a wide scope, it is essentially a unified system of opposites (human, calculation environment) of subjectivity and objectivity, perception and rationality, intentionality and formalization; that is all. Its core value is still inseparable from breakthroughs in basic theories, rather than data, algorithms, computing power, and experiments. The human-machine hybrid is not only about creating more advanced machines, designing better algorithms, and obtaining more data, but also about the transformation of a human's own intellectual nature, that is, the transformation, reshaping, and change of thinking logic.

In the vast universe, human beings are as small as a drop in the ocean, just like an element in a huge matrix. There is no way to know where the limits of artificial intelligence will be in the future, and whether human-machine hybrid intelligence will enable machines to break through the bottleneck of self-awareness; it is too early to draw a conclusion.

6. Persona: the bridge for human-machine communication

6.1 Introduction to the concept of persona

Persona is a concept first introduced in 1983 by Alan Cooper, the American software designer and "father of interaction design" and proposed in 1998 in his book *The Inmates are Running the Asylum.*

Ogilvy and Mather used a similar concept in marketing in 1997, known as "Customer Prints" which is a categorical description of the typical customer in everyday life.

> *Every strong brand has a group who share the brand's values. The population as a whole is divided into different groups according to their typicality, each of which has the same or similar buying behavior, and the personality and characteristics of the group owned by a certain brand (product or service) can be understood in terms of shared values, attitudes and assumptions. Customer Prints are the description of the essence of life of these different customer groups.*

Persona plays an important role in the context-based design approach, as it can be used repeatedly in the framework definition phase to generate design concepts and in the optimization phase to provide feedback to ensure correctness and consistency of the design.

6.2 Human and machine

Objectively speaking, artificial intelligence is only a describable and programmable part of human intelligence, while human intelligence is the product of the interaction between humans, machines (object), and environmental systems. The mechanism of intelligence generation, in short, is the result of the superposition of the interaction of human-machine-environment system (machine is a man-made object), and it is composed of the state, situation, and potential (referred to trend) that are derived from the superposition of the changing states of human, machine, and various environments. The state of the three changes is different, either good or bad, high or low, prosperous or inverse, and the generation of embodied intelligence is determined by the degree of harmony and resonance of the system state and trend. There is a constructive and destructive interference effect among these three sides, which can be enhanced or eliminated. The intelligence will be strong when the three coincide and will be weak otherwise. How to tune the common frequency of the three is the key to human-machine hybrid intelligence. Contemporary artificial intelligence has developed from the initial completely artificially compiled machine automation to artificially precompiled machine learning, and the next development may be to achieve machine cognition and eventually machine awakening through the method of human-machine hybrid intelligence.

Since humans are the main subject of the human-machine-environment system, only by deeply understanding the operational characteristics of humans in the system can we develop a high-quality and efficient system that maximizes the overall capability of humans and human-machine systems.

In the human-machine-environment system, the human being, as the dominant power, is closely related to the development, analysis, and performance of the system. Although the human body is material, it has sentiment, thinking, and intelligence. A good human-machine-environment system must be built on the basis that the machine has good recognition for the human.

So, how does a machine understand a human? How does one build the communication between the two?

The question is equivalent to how does one perform in-depth persona? Persona translates the knowledge, values, and emotions of the user into computational models that can be understood by the machine, enabling the computer to understand how the user perceives outside world, the process of recognizing things, and the patterns of behavior. Based on the understanding of the user, the machine can further change its own behavior pattern to adapt to the sensory style of humans, further enabling the human-machine interaction to adapt to human characteristics as much as possible according to the user's needs, forming the most harmonious human-machine integration.

Persona is a technology for studying deep situation awareness systems based on human behavioral characteristics, that is, the perception and response capabilities of human organization in uncertain and dynamic environments. It is an important reference for emergency command and organization systems in social systems (war, natural disaster, financial crisis, etc.), rapid fault handling in complex industrial systems, system reconfiguration and repair, design and management of humanoid robots in complex environments, etc.

6.3 Persona based on deep situation awareness theoretical model

Deep situation awareness is the perception of perception, which is based on human cognition, preference, habit, emotion, memory, and perception, coupled with machine computing coordination, remedying with each other and learning from each other. The construction of persona is the process of generating trend from state. The persona is a comprehensive manifestation of a series of human cognitive activities (such as purpose, sensation, attention, motivation, prediction, automaticity, decision-making, motivation, experience, and knowledge). Based on the theory of deep situation awareness, persona can deeply portray people in four dimensions: natural attribute, value orientation, behavior habit, and cognitive characteristic.

In fact, replacing the concept of "purposive behavior" which particularly belongs to living beings, with the concept of "feedback" and considering the working behavior of machines designed according to the feedback principle as a purposive behavior, does not break through the conceptual barrier between living beings (people) and nonliving beings (machines). The reason is simple: human "purposive behavior" is divided into simple explicit and complex implicit. The former can be approximately equivalent to the "feedback" of nonliving machines. "Purposive behavior" can be approximated by "feedback" from nonliving machines (stimulus-response), while the latter—intentionality—is far from being approximated by "feedback" because this intentionality can be delayed, increased or decreased, and concentrated. The definition of "reflection" is more accurate, but the concept of "reflection" is hardly comparable to that given by the inanimate machine (stimulus–choice–reaction) with nonliving machines. The purposefulness of "reflection" can be characterized by subjective valence, which will be another key to the human-machine hybrid. Value consists of both attractors and motives. Reflection is an unproductive feedback or an organized feedback. Autonomy is organizing adaptability or organized adaptability. Accordingly, we combine Endsley's three-level model of situation awareness with Wiener's idea of "feedback" and propose a model of deep situation awareness based on "feedback."

1. State

The "state" in deep situation awareness is defined as all kinds of subjective and objective data that characterize the state of an individual in the human-machine-environment system. Persona extracts any useful information and knowledge from a large amount of information. The construction of natural attribute dimension in persona is the basis, which corresponds to the "state" of deep situation awareness. The natural condition dimension of the persona contains relatively static demographic information, physical conditions, and social circumstance. The information of the natural condition dimension is the most accessible apparent information, and although the process of constructing persona tends to weaken this information, it is undeniable that natural conditions are one of the foundations of user perception and behavior.

2. Trend

"Trend" is defined as the developing trend of an event, which corresponds to the value dimension in the persona. The value orientation dimension contains an individual's expectation of the value one can produce for himself or herself, for the group, and for society. Humanistic psychology views "self" as the psychologic motivation for behavior and considers the difference between the ideal self and the real self as the driving force of behavior. There is a gap between the real self and the desired self, and bridging this gap is the realization of self-worth, the fulfillment of expectations. This gap bridging is the way to achieve happiness, which is to move closer to the desired self. Value orientation is an implicit or explicit view of what is valuable, a characteristic attribute of individuals and groups that influences people's choice of behaviors, means, and ends. Individual values guide and drive a person's decisions, performing on his or her behavioral norms and beliefs. People's ultimate thoughts and explorations of purpose, existence, and meaning are stored in individual values in various forms.

The most significant meaning of the value orientation dimension for the construction of persona is to identify the source of motivation that influences user behavior and provides a directional vector force for user decisions and behaviors. The value orientation dimension reflects on the deep needs of users and expresses users' expectations of themselves.

3. Sense

In general, sense corresponds to fragmented attributes, while percept is the simultaneous establishment of relationships. "Sense" is defined as the awareness of the "state" of the system. Just as "sense" corresponds to deep situation awareness, behavioral patterns are a very important part of user research, which are also the backbone for building persona models. Behavioral habits embody the four dimensions of representation and are therefore easily observed, allowing for problem identification, optimization, and iteration.

4. Percept

"Percept" is defined as the understanding of "trend" The interaction between "sense" and "percept" is closely related. Of the persona, the relationship between behavior and cognition is also close. "Percept" includes sensation, perception, memory, and personality, which shows the user's energy tendency, information acquisition, decision-making, and lifestyle characteristics. It corresponds to people's expectations of how to use the product or how to achieve specific interaction behaviors, and this dimension focuses on the user's five senses (especially audio and visual), interaction, and the physical design of the product. Cognition influences behavior, and the results of behavior also influence cognition. A person's behavior is the external form of his or her cognition in a certain social and cultural space. To build a persona is to outline a typical user's lifestyle in a focused manner.

5. Reflection

Reflection is the process by which users think and process their own experiences or the information they receive, and it influences their aesthetic tendencies, preferences, and values. It has a deep and lasting motivation that affects the user's decisions, even if the user does not necessarily appreciate the underlying reasons for the decision. The reflective dimension is strongly influenced by the value orientation dimension, which largely affects the user's thinking and opinions about things or events.

Natural attributes are the basis for constructing dynamic persona, while value orientation, due to its far-reaching influence on user lifestyles, has a long-term impact and serves as a solidifying layer that influences user behavior and cognition together with natural attributes. Users' behaviors and habits are the externalized expression under the combined effect of natural attributes, value orientation, and cognitive characteristics. Behavioral habits and cognitive traits have an intrinsic relationship that influences each other.

Natural attributes, values, behaviors, and cognitive traits are not independent planes. They are mutually driving, interacting, and influencing each other. Therefore, considering user behavior patterns alone is a one-sided approach to constructing persona, which is also the reason why traditional persona tools are found to deviate greatly from the constructed persona in the process of using. Perhaps by starting from these four dimensions, the constructed persona can more accurately describe typical users.

CHAPTER 11

Human wisdom and artificial intelligence

Contents

Up to now, the storage of machines has still been formalized, while human wisdom is often visualized. The artificial intelligence (AI) calculates through formalization, while human beings suanji through the integration of objective logic and subjective intuition. The calculated predictions do not affect the results, but the suanji expectations often change the future. In a sense, the deep situation awareness is not perceived by calculation but cognition. Autonomy, an attempt to modify from the inside out, has both advantages and disadvantages. It is the verification of experience and the analogic transfer of experience. If computing displays the function of brain, then suanji displays that of the mind, defining the size of the world.

Some people think that AI is a way for human beings to know about and know themselves. In fact, AI is just an attempt by human beings to understand themselves because the coordinate origin of "who am I" is far from being determined. The question of "who am I," the initial problem of autonomy, is the origin of the coordinate system of human intelligence. And memory is a directional consciousness vector (intentionality) in this coordinate system, which is different from the storage of John von Neumann's computer system. The program rules and data information are not static but changing with the system interaction in the human-machine hybrid environment (so the brain-like significance alone is not highly significant). The degree of flexibility of this change often reflects the degree of autonomy. For example, language communication, a model of autonomy, is carried out according to the interactive situation (not the scene). No

Integrated Human-Machine Intelligence
ISBN 978-0-323-99562-7
https://doi.org/10.1016/B978-0-323-99562-7.00011-5

matter how it is tested, it is a reflection of script and nonscript, and its accuracy can be used to determine whether man-machine is right or wrong. It is said that language has three references, namely, references to name, mind, and objects, and that the study of these three and their relationship has always been a difficult problem and challenge for AI. Coincidentally, in the 19th century, British scholars put forward the concept of signifier and the signified. If you think about it carefully, I am afraid these are nothing more than the attributes of things (signifier, feeling) and the relationship between them (signified, perception). In fact, a word, a sentence, a paragraph is inseparable from the autonomous contextual limitation, and what we know (signified) is much more than what we can say (signifier). If you do not believe it, think of the people you have seen who can talk with their eyes. The rooted cause lies in transforming from sensibility to rationality: sensibility is the wormhole of rationality, passing through its bondage and restraint; rationality is the black hole of sensibility, limiting its willfulness and arbitrariness. It can be said that the independent consciousness controls reason and is enslaved by reason at the same time.

At present, the most challenging problem in the field of intelligence is the degree of difference between people's intentionality and behavior. Behavior can be objectively explicit, while intentionality is subjectively implicit. Intentionality includes reflection and imagination: reflection is the summary of experience, that is, prethinking, while imagination is the assumption of the future, that is hindsight and meditation, both aiming to solve current problems.

The difference between formalization and intentionality lies not only in the difference between the exterior and interior, but also in the difference between phenomena and rules. For example, the sunset and pains in leg. The former is formal: the phenomenon is the setting of the sun, which is actually a concrete manifestation of the objective law that the earth rotates around the sun. The phenomenon of the latter is leg pain, which is essentially neuralgia in specific parts of the brain. Such as sunset phenomenon, the physical mathematical sense of time and space and positioning are very different from those of the intentionality. Leg pain, the pain of state space and fact space, is inconsistent with the suffering of potential space and value space.

Intentionality, the subject's perception of things, is inherent and individualized. Formalization is the perception of objects, such as physical theorems and mathematical formulas, which are divorced from individual existence and accepted by most people, so they are external and common perception.

The essence of intelligence lies in the judgment of autonomy and "similarity," and in properly grasping the "similarity benchmark". One of the advantages of people over machines is that they can discover patterns of things earlier from less data. One of the reasons is that the machine does not have a coordinate origin, that is, the question of who "I" am. For human beings, things are nonexistent being. Existence is not objective, but the result of our observation with subjective purposes, and this mixture of the subjective and the objective is often the product of the context of the situation. For example, the construction and deconstruction of the process of being, should, want, can, and change are often carried out at the same time. In addition, even the same sense, such as looking, also has concrete direction and abstract meaning, and handshake, a kind of physical contact, can be accompanied by psychologic cues. When the human brain carries out autonomous activities, it can produce "mapping from Euclidean space to topological space". That is to say, when selecting and controlling, people can respond differently according to specific purposes and the changing similarity benchmark (not the proximity in the Euclidean space, but the reasonable connection network).

Compared with the machine, humans have strong language or information chunking ability, though limited memory and rationality; the machine has weak language or information chunking ability, yet unlimited memory and rationality. The simultaneous realization of its language (program) operation and self-monitoring mechanism should be the basic principle to ensure the reliability of the machine. People can communicate without considering grammar when using their mother tongue and can perceive the polysemy of language, pictures, and music in many situations. For example, people's auditory, visual, and tactile senses are discriminative and emotional. We can often perceive information or concepts that cannot be expressed (such as thinking about philosophy, which is difficult to obtain through learning). Although the machine can play chess and answer questions, it is weak in adapting to different domains, reflecting contradictory or ambiguous information (lack of necessary competitive risk selection mechanism), selecting the primary from others, and comprehensively discriminating. It cannot form concepts and put forward new concepts through induction, reasoning, and deduction, let alone produce metaphysical theories.

Master's learning methods: first to select through sense, and then to think through reasoning. Sensibility is the internal intentionality and the correlation manifestation of potential factors. Rationality is a formal intentionality and an explicit correlation.

In addition, the difference between human learning and machine learning is that human learns fragmentally and integrally, so human learning has strong adaptability. It has been carrying out the stable prediction and unstable control in the situation of insufficient information. The scenes out of prediction and out of control occur from time to time. It becomes more and more necessary to provide multiple timely feedback adjustments and corrections. In this regard, human beings' processing mechanism in unstructured and nonstandard situations is better than machines, while machines are relatively better than people in structured and standardized situations. Moreover, this adaptability is cumulative and will gradually form an individualized and reasonable expectation. So far, the autonomous (expectation + prediction + control) mechanism has begun to emerge and grow. Intelligence is not an encyclopedia but contains a lot of fiction and imagination. Einstein said that "Imagination is more important than knowledge, because knowledge is limited, and imagination summarizes everything in the world, promotes progress, and is the source of knowledge evolution". Fiction is the essential characterization of intelligence, represented by what can be understood but cannot be described, such as something that you seem to have met before, something that looks right but in fact wrong, and something that looks wrong but in fact right.

The mainstream method of machine learning is to generate a "model" from the samples through a "learning algorithm" and then to solve practical problems through the model. Actually it is not strictly distinguished between the learning process and problem-solving process, but it decomposes the whole operation into a large number of "basic steps," with each step being realized by a simple algorithm and a reasoning rule. The connection of these steps is determined in real time and is generally not strictly repeatable (because the internal and external environment are not repeatable). Therefore, a general intelligent system should have no fixed learning algorithm and no constant problem-solving algorithm, and "learning" and "reasoning" should be the same process.

In addition, human learning is the integration of causality, correlation, and even customs. Some of these can be programmed, but most are still difficult to describe clearly, such as some subjective feelings and tacit knowledge, whereas the explicit knowledge connotation of machine learning is far greater than its implicit concept denotation. In fact, for people's cognitive process, the relationship between rules and probability is cohesive. Rules are the existence of large probability, and the essence of probability is the state of not forming rules. Habit is a regular unconscious behavior, and learning is a cumulative process of probability, including familiar analogy and unfamiliar correction. Generally speaking, the former is unconscious, and the latter, a composite process, is conscious. In addition, the speed of human's information processing is varying, sometimes automatic release of subconscious habit, sometimes semiautomatic balance between consciousness and unconsciousness, and sometimes purely, leisurely manual. However, it is not a simple process of information expression and transmission but also a process to construct and organize corresponding grammatical states in the knowledge vector space and reconstruct various semantics and pragmatic systems.

Moreover, the freely regulated environmental system triggers the reverse movement of the autonomous system, resulting in multidirection or multimovement between man-machine and the environment, which leads to contradictions and conflicts. The solution of this inconsistent or even opposite problem is often beyond the power of simple mathematical knowledge. A problem is a mathematical discussion when it is solved with boundary, condition, and constraint. When the same problem is solved without boundary, condition, or constraint, it often becomes a philosophical research. For example, how does fiction correct reality, and how does reality feed back fiction? This will be a very interesting problem.

1. Intelligent thinking

It is always believed that AI is only a descriptive and programmable part of human intelligence, and human intelligence is the product of the interaction of human beings, machine (object), and environmental system. In short, the mechanism for generating intelligence is the superposition result of the interaction of human, object (machine is a man-made object), and environmental system, which is composed of the situation, complexion, and trend (hereinafter referred to as trend) derived from the superposition of the changing states of human, machine, and various environments. The states of the three are sometimes good and sometimes bad, high and low, positive and negative. The generation of intelligence is determined by the harmonious resonance between state and trend of human-machine-environment system. The three have constructive and destructive interference effects, enhancing or eliminating intelligence. Three in resonance will generate strong intelligence, while three not in one generate weak intelligence. How to achieve the resonance among the three is the key to human-machine hybrid intelligence.

In poems, we can often find a kind of illogical logic. Just like autonomy, it is with purposeful creativity, including both assimilation mechanism and adaptation mechanism. All logic contains nonlogic, just as all nonlogic contains logic. There are intentions in the forms, and forms in the intentions; there is calculation in suanji and suanji in calculation. This "intelligence" in good poems is stored in the emotion of words and their semantics. The uncertainty of poems as an intelligent agent is caused by the variability of characterization and reasoning. Two hypotheses are hidden behind its mechanism: program variability and description variability. These two are also one of the reasons for the inconsistency between expectation and reality. Program variability indicates the difference in the derivation of prospect and behavior, while descriptive variability is a dynamic nonessential characterization of things. Schiller called aesthetic consciousness "game impulse," and "game impulse" is the unity of "perceptual impulse" and "rational impulse". Simple "perceptual impulse" makes people limited by perceptual material desire, which is a kind of nonfreedom; the simple "rational impulse" makes people limited by rationality (including moral obligations), which is also a kind of nonfreedom. Only the "game impulse" combining the two can surpass the limited and reach the infinite; then there can be freedom without any restriction.

Einstein believed that the development of Western science is based on two outstanding achievements, that is, the formal logic system invented by Greek philosophers and the possibility of finding causality through systematic experiments. In fact, the development of Oriental thought is also based on two outstanding achievements: one is the virtual and real analogy change system of the *Book of Changes*, and the other is the value relationship between one and more in *Tao Te Ching*.

Some people think that the AI system is a tool just for one-time use, rather than intelligence that can be used repeatedly by humans. Judea Pearl, the Turing prize winner and the father of the Bayesian network, once mocked himself as a "rebel of AI Community," because his view on the development of AI is contrary to the mainstream. Judea pearl believes that despite significant progress made in the existing machine learning models, it is a pity that all models are just accurate curve fitting of data. From this point of view, the existing models only improve the performance based on the previous generation, and there is no progress in the basic ideas. For example, machine learning is to find eigenvalues hierarchically. The quality and quantity of input labels are essential. Human beings usually know the connotation and denotation of each label and the bivergentum relationship between them, which machines do not understand. Machine learning, just stipulated dance movements of symbols, mechanical segmentation of scenes, cannot produce overall feeling, perception, or understanding. Understanding is seeing the connection. Among all kinds of intelligence, the flexibility of the input terminal is significant. The bivergentum of the connotation and denotation of each label and the speed have largely determined the quality of intelligence (like many people, ideals and dreams in childhood determine the direction and height of their life). At the same time, it is also the basis for automatically generating new information and knowledge, new functions (functions), new networks and new capabilities. Generally speaking, if the range of markers is too large, it is difficult to store; while if it is too small, it is incapable to express the meaning. The balance between the uniqueness and generalization (i.e., bivergentum effect) of markers naming shall be achieved, which is to achieve the unity of opposites between comprehensiveness (semantics) and centralization (uniqueness). Therefore, the minimum granularity of input data, information, and knowledge identification, i.e., the size of boundary range, is significant, If the particle size threshold is too large, it is easy to cause intelligent uncertainty. On the contrary, if it is too small, it is easy to lose key features. These will directly affect the intelligent architecture: characterization-identification-networking-optimization-correction-iteration.

In addition to the random changes in the characterization of the input, reasoning and decision-making are also very important. If the iterative relationship between knowledge-driven and network reconstruction is not handled well, it is difficult to generate new knowledge from intelligent content. For intelligence, planning is "just a basis for deliberation." An excellent intelligent system relies not only on logic but also on intuition. It is more faithful to its innovation inspiration than the promotion of rigid rational logic. Hayao Miyazaki said, "The so-called film does not exist in his mind, but in the space above his head." You cannot rely on logic to make a movie, or if you look at it from another perspective, anyone can make movies with logic, but my way is not logical. I try to tap my subconscious. The spring of thinking is opened at some point in the process, and all kinds of views and ideas rush out." Although many commentators" have pointed out the structural imbalance of Hayao Miyazaki's works, few have criticized them. On the contrary, his weakness is regarded as his own style:

"By feeling, Hayao Miyazaki can create films satisfying the audience, reflecting his extraordinary talent."

Compared with machine intelligence, human intelligence has many different mechanisms. The mechanisms of reflection, meditation, and confession are fundamental. Reflection is a nonfactual inference and a repetition of various assumptions. It can deduce "doing one thing" into "doing many things," which is also beyond the reach of the "feedback" mechanism in machine intelligence. Meditation is a relevant and irrelevant nonlogic. It regards time, space, and logic as nothing. It is a bit like some good poems, which shuttle willfully among movement, stillness, nothingness, emptiness and reality, strength and weakness. Confession is to format yourself. Everything has its advantages and disadvantages. Over time, cleaning is inevitable. The more garbage information is deleted, the less blocking, the less complexity, the more intelligent it will become, and the easier it is to form wisdom.

The three philosophers are imperceptibly following the law of innate and postnatal research fields: from logic to non-logic. There are two reasons. First, in terms of the research process, "logic" mainly involves "judgment theory," a kind of high-level consciousness activity. Secondly, it involves the universal validity of logic and mathematics. Frege pointed out that psychologic law is an empirical law, which is occasionally true, and that the laws of logic and mathematics have universal inevitability, which cannot be explained by the laws of experience. This may be the breakthrough and entry point in the field of intelligence research.

"Logic" also involves reasoning, another kind of higher-level conscious activity. To clarify the judgment theory, it is necessary to study lower-level conscious activities such as feeling and perception. In particular, it is necessary to conduct an in-depth study on the bivergentum characterization mechanisms of physical data and psychologic information. The "name" in "The name that can be named is not the enduring and unchanging name" mainly refers to the dynamic characterization, naming, definition, and categorization. While the "Tao" in "The Tao that can be trodden is not the enduring and unchanging Tao" reflects the integration of facts and values, things and relations, including both objective being and subjective "should" It not only involves the universal validity of logic and mathematics, but it also covers occasional truth of psychologic law. It is a collection of logic and nonlogic. Emotion may be a complex of one's own logic and others' nonlogic.

Lao Tzu's Tao is extremely natural. For example, when he said, "Wisdom comes out, there is great hypocrite," it means that the more people pursue wisdom, the more man-made things there are, and the greater is the element of self-righteousness. This needs to reverse the usual view of the world and let things see us, just like the compound eyes of insects. Cezanne said, "A good painter looks at the world not from the outside but from the inside." Intelligent people often look at the world in the opposite way to their daily life, such as Sai Weng in the story of Sai Weng's Horse Was Lost and the spark on Jinggang Mountain. We should not just look at the world in a scientific way. AI and its philosophy should make people really open to nature and make the human-machine-environment system speak and display to us in its own form.

The idea of human-first is understandable in human-machine hybrid intelligence at the primary stage, in which the machine is mainly an auxiliary. However, with continuously grading functions of the machine, especially when there are faint signs of human-like ability, the concept of "human-first" may be shaken. We can see that from the enlightenment in *Tao Te Ching*. First, we should distinguish the existence of Tao itself from man-made structure. Second, to overcome the situation of human-first, we need to avoid human efforts. But to overcome these human efforts, another kind of human effort is needed (a bit like Gödel's incompleteness theorems). Lao Tzu mainly wants to overcome humanitarianism, which is highly viewed by Confucianism, in which man is centered while nature is discarded. Lao Tzu believes that man is not important, and that the Tao existing in nature is the center. Man is only a member of nature and a kind of thing. There are endless changes in the infinite universe. Only people who follow it can give better play to their subjective and objective initiative. For example, from the Renaissance to the industrial revolution and later, the world image has been updated at regular intervals. In the West, each stage will produce a new view of the world.

The most important thing that Western science and technology has learned from Oriental thought is how to overcome subject metaphysics. Many people believe that subjectivity is associated with modern science and technology. For human beings, modern science and technology is just a kind of machine. It controls nature with more and stronger forces. This process is endless. The will to power in the Western world is to overcome metaphysics with a stronger thing and surpass it with a stronger force. This approach is bound to create a vicious circle: you are no different from the strong if you use the stronger to overcome the strong. Oriental philosophies provide a different way: overcoming the strong with the weak rather than the stronger. Water is soft and weak, "yet nothing is stronger than water when it comes to breaking something strong." Lao Tzu believes that female, negative, and passive things can overcome male, positive, and active things. Heidegger once summed it up in one phrase: verbindung der metaphysik, which means to overcome the superior through the inferior. The Eastern thought of "overcoming hardness with softness" has derived "judo." This "softness" in judo is not the weak in the general sense, but it uses your own strength to defeat you. It is to gear the offensive and linear force to the attacker himself and to defeat the attacker.

The reason why it is easier to generate groundbreaking ideas in humanistic art than in science and technology lies in the fact that the former pursues subjective value and significance rather than simple objective facts. Therefore, humanistic art, philosophy, and religion provide people with a broader space to imagine. It can be said that what people see is not important, but how people interpret what they see. The essence of emotion is the judgment of value. It is very difficult to quantify value, which requires a clear understanding of the essence of value and the essence of calculation before value calculation can be done. Some people think that "if it is not easy to calculate absolute value, it is OK to calculate relative value." In fact, it is more difficult to calculate relative value because factors are changing; even the coordinate system is changing.

In essence, value and meaning change with the interaction between subject, object, and environment. For example, it is possible that a piece of information has value one second, and has no value at all the next second, or it may have greater value the next second. The relationship between things changes quickly, and the value changes greatly in the process. How can this be realized by algorithm? Concerning theories remain unclear. Here I still want to repeat Derrida's famous saying, "Give up all depth, and appearance is everything". By saying so, he means that life itself does observe logic: it is illogical and nonstandard, just like philology, dancing with a strange logic.

With the development of mankind and the society, people's understanding of war, game, and confrontation has also changed greatly. For example, an increasing number of people believe that conquest is greater than destruction. And people's understanding of intelligent thinking and consciousness is also changing with the times. From this perspective, the future command and control system may be the unity of art and technology and the resonance of future art and next generation theory, and the relations between qualitative/quantitative, local/overall, dispersion/aggregation, primary/secondary, time/space, logical/illogical, confrontation/compromise, coordination/independence, and even between the army and the people may change a lot. Great changes will also take place in the border. But no matter how advanced technology is and how powerful equipment is, they are to serve people in a better way. And there will be no fundamental change in the concept of "human-first" in a short time.

Objectively speaking, complexity science is a wrong concept. Complexity is a process of multidisciplinary integration, while science is a process of "learning by branches." The former is an aggregation process, while the latter a diffusion process. The former is positive while the latter negative. Therefore, the correct title should be the field of complexity research. AI is a prominent representative in the field of complexity research. It is not a science of "learning by branches," but the integration of multidisciplinary complexity.

From different perspectives, perhaps the objective law is not unique. It can be said that "it is a range viewed in face and peaks viewed from the side, assuming different shapes viewed from far and wide." A thing is not viewed as the same from the perspectives of physics, mathematics, psychology, game theory, ethics, and management. Perhaps each has its own objective laws and logical clues. The so-called nonlogic is often just the twists and turns, sewing and connections at the right time and in the right place of these various logical clues. The truth is that logic serves nonlogic and intelligence serves wisdom. Specifically, there are both logical and nonlogical states, trends, senses, and percept in deep situation awareness in human-machine hybrid intelligence. Logical states can be formally calculated, and nonlogical ones should be calculated intentionally. At present, AI promoters are eager to figure out the former, but they generally turn a blind eye to the latter because this problem is so challenging that it affects their earning, consumption, and the normal use of IQ and EQ.

Being politically sound and professionally competent, which was once the standard of employment in China, has gradually been paid more and more attention. Why? It is very simple. People cannot stand without faith and morality. If one has no faith, the greater his talent is, the greater is the harm. Ideological quality, ethics, and morality are essential, while talent, learning, and intelligence are small in process. It can be said that the slightest mistake is a thousand miles away. At present, many people intentionally or unintentionally regard informatization and automation as intellectualization rather than "wisdomization." Why? Because it is lack of "virtue." Wisdom is the intelligence integrated with Tao, virtue, ethics, and reason. How dare you talk about wisdom without benevolence, righteousness, and morality. Benevolence and understanding, significance is response; Tao is the way; morality is equal to gain. Benevolence, righteousness, and morality mean that "people gain through the right way." This is also the core and driver of wisdom: intelligence and ethics. Ethics mainly involves the norms of thought and behavior jointly recognized among people and between people and society. It is the classification, analysis, and decomposition of facts and things in a certain group of people. It involves the problems concerning value: security, privacy, prejudice, substitution, and inequality. It is also the humanitarian basis for distinguishing the right from wrong and producing compassion and empathy. However, the machine, human beings' formalized information characterization, is still helpless in solving the aforementioned problem.

Wisdom is the combination of objective "being" and subjective "should." It bridges the gap between fact and value, the subjective and the objective, ethics and intelligence. It is not only an important symbol in the field of general intelligence or strong intelligence, but also the embodiment of reflective intelligence and insight intelligence. Many truths in the world may be round. When you think they are right, they may become wrong slowly or suddenly, and vice versa. The reason why wisdom is more important and smarter than intelligence is that it may foresee unexpected things that cannot be told by intelligence.

A field is often based on a mistake, but this is not necessarily fatal. In fact, many fields are based on mistakes. For example, chemistry is based on alchemy, and cognitive science is based on the erroneous assumption that "the brain is a digitalized computer." Similarly, AI starts with the error of logical reasoning.

The programmable part of human intelligence has been transformed into AI, while many declarable but nonprogrammable parts and more nondeclarable parts of human intelligence are far from being transformed, particularly nonlogical parts that cannot be explained and learned (such as the deep level of emotion, the random change of consciousness, and the fuzziness of thinking).

Human beings have a more abstract ability that has not yet been discovered, which is the intelligent system starting from nonlogic. This intelligence may start with mistakes, but end with the right one; it may start with chaos, but end with faint order; it may start with facts, but end with the value of intention. On the contrary, it may also start with the right, but end with kinds of mistakes. It may start with order, but end with all kinds of chaos. It may start with value, but end with relevant facts. It can be said that people always start from ignorance, and later, as Michael Polanyi said, "the more we know, the more needs to learn."

In short, 5G means fast communication and AI means accurate processing. The combination of the two will bring faster and more accurate interaction between human, machine, and environment, which in turn will affect all aspects of production, business, and life and will greatly change and reshape the fields of clothing, food, housing, transportation, and entertainment. It fully reflects the strength and flexibility of logic serving nonlogic.

In view of the aforementioned understanding, some friends even mistakenly think, "People, as soon as they are born, are counting down! Fortunately, they are with 5G and AI." It seems that the combination of the two will greatly contribute to the coming of revolution and innovation, and then come unimaginable, dizzying, dreamlike situations. And Cinderella will no longer be gray, Smurfs will no longer be blue, Romeo and Juliet will no longer be sad, and Liang Shanbo and Zhu Yingtai will have a happy life together.

In the East, sensibility was once regarded as an out-of-proportion relationship between all things, not much like geometry. In the West, the element of Newtonian ontology is substance, which exists in absolute time and space. In the Newtonian system, there is no need to introduce consciousness, life, organization, or purpose, which are the important points of Oriental thoughts.

5G and AI, so fast and accurate, will bring more benefits, but they will also bring many unexpected problems. If you think about it carefully, many problems do not necessarily need to be solved quickly and accurately. For example, many decision-making problems need to be solved slowly, while others need to be vaguely solved. In addition, although 1 and 1 means 2 in modern mathematical logic world is true, in nonlogical environments such as psychology, management, and ethics, 1 and 1 does not necessarily mean 2. Some people even think that in the world dominated by hearing and smell (rather than vision), addition, subtraction, multiplication, and division may not be tenable at all. The problems of relativity, immeasurability, incompleteness, and impossibility in the man-machine-environment system still exist in 5G and AI.

A man-machine system often exists in a changing environment, in which the attributes of and relationships between man, machine, and environment will also change according to the situation, and the consciousness, life, organization, and purpose involved will fluctuate at any time and with the situation, which may weaken the advantages of 5G and AI. How do we integrate the ethics and morality into a future science and technology combination (including 5G and AI)? This has become a difficult and hot point for in-depth thinking of current science and technology innovation.

The more open the system is, the more it can handle high complexity, and the stronger its ability to mutate and produce new order. Mr. Zhu Songchun wrote an article, introducing two paradigms of AI. One is called "parrot paradigm." Parrots can talk to humans, but they do not understand what you are talking about. For example, when you say Lin Daiyu, a parrot also says Lin Daiyu, but it does not know who Lin Daiyu is. The other is a "crow paradigm." After finding walnuts, crows will throw walnuts on the road, let the car crush them, and then eat them. But because there were too many cars on the road, the crow could not eat walnuts, so the crow throws the walnuts to the zebra crossing because there are traffic lights there. When the green light is on, the cars stop, and it can eat the walnuts. This is an amazing example, because crow has neither big data nor supervised learning, but it can independently study the causality and then use resources to complete the task. The power consumption is very small, less than 1 W, which gives his research team great inspiration.

In fact, there is also a "crow + parrot" paradigm, which can be either top-down or bottom-up. For example, mobile communication is a kind of two-way work, transmitting data at the same time. There are two ways to realize two-way operation. One is Frequency Division Duplex (FDD), which performs input and output (transceiver) based on two frequencies. The other is Time Division Duplex (TDD), which is based on one frequency but processing at a different time.

In man-machine hybrid intelligence, deep situation awareness is a multidirectional working mechanism between man, machine, and task environment. It not only includes transmission of data, information, and knowledge, but also transmission of intelligence and wisdom (including ethical intelligence).What should be done to achieve effective multidirectional work? How does data become information? How does information become knowledge? How does knowledge become intelligence? How does intelligence become wisdom? How does wisdom become ethics? To distinguish the right from the wrong (gaining wisdom) is the most difficult because there is a great amount of plausibility, such as the rotating yin-yang eight trigrams: existence, reality, movement, and stillness.

2. Human consciousness has a purpose

John Rogers Searle, born on July 31, 1932, in Denver, USA, is a professor of philosophy whose main research is the "purposefulness" of language. He rejected the notion of "strong AI," arguing that consciousness is an emergent phenomenon of an organism. Minds have intentionality while computers do not, so they have no consciousness.

He discovered the property of an intentional phenomenon and called it "direction of fit". For example, when one sees a flower, their mental state is to fit the world. The direction of fit is mind-to-world. But if that person reaches out to pick the flower, they are aiming make the world fit with their own mental state. So the direction of fit is world-to-mind. He also develops the term "background," used here in a rather technical way, which has been the source of some philosophical discussion. Roughly speaking, it is the context within which the intentional act occurs. Importantly it includes the actor's understanding of the world, including that others can and do participate in intentional activities.

He provides a strong theoretical basis for the use of the notion of intentionality in a society context and raised the following five theses.

(1) Collective intentional behavior exists and is not the same as the summation of individual intentional behavior.

(2) Collective intentions cannot be reduced to individual intentions.

(3) The preceding two theses are consistent with two constraints:

① Society consists of nothing but individuals; there is no such thing as group mind or group consciousness.

② Individual or group intentionality is independent of the truth or falsehood of the beliefs of the individual.

To satisfy these theses, Searle develops a notation for collective intentionality that links an individual intention with a collective one but keeps the two types of intentions distinct. In effect, an individual intention can have as its outcome a collective intention. Forming a collective intention presupposes that one understands that others can participate in the intention. Therefore, we have the following:

(4) Collective intentionality presupposes a background sense of the other as a social actor, as being able to participate in collective activities.

Together, these theses lead to the following claim:

(5) The theory of intentionality, together with the notion of a background, is able to explain collective intentionality.

Searle has more recently applied his analysis of intentionality to social constructs. His interest is in the way in which certain aspects of our world come into being as a result of the combined intentionality of those who make use of them. For example, a five dollar note is a five dollar note only in virtue of collective intentionality. It is only because I think it is worth five dollars and you think it is worth five dollars that it can perform its economic function. This is so despite the apparent role of the government in backing up the value of its currency. Imagine a case in which you were attempting to make a purchase from someone who did not recognize the value of the note. Until you can convince them of its value, all you have is a colored piece of paper. Such socially constructed objects permeate our lives. The language we use, ownership of property, and relations with others depend fundamentally on such implicit intentionality. Searle extends his analysis of social reality to the creation of institutions such as marriage and universities. He claims that the value of the five dollar note and the institution of a university are created by the function of three fundamental primitives: collective intentionality, the assignment of function, and constitutive rules.

Searle's approach to social construction is quite distinct and divergent from those who would suggest that there is no such thing as a mind-independent reality, that what we call reality is a social construct.

In the long history of humankind, the ancient Egyptian pictogram, the ancient Babylonian Cuneiform, the seal script in ancient Indus Valley, and Chinese oracle bone inscriptions have formed the world's four major ancient writing systems. However, only Chinese oracle bone inscriptions have traveled through time and are still in use and full of vitality. The fundamental reason lies in the appearance of the official script (Li), a kind of ideographic character current in the Han Dynasty (206 B.C. to 220 A.D.). Since then, the Chinese characters have developed from being pictographic to being ideographic, becoming an incessant civilization. The core and key of intelligence science is still to complete the breathtaking jump of "getting the idea while discarding the form." At present, various signs of scientific and technologic progress indicate that the combination of man's intentionality and the machine's formalization is possibly the highest form to achieve intelligence: "the most possible way to get the meaning while discarding the form."

Judea Pearl, a pioneer in AI and the father of Bayesian network, says AI is mired in probabilistic correlations, ignoring causation. Judea Pearl thinks researchers should study causation, a possible path to a brilliant machine.

Today, the essence of a human-machine hybrid intelligence system is "things are different from people." Machines can handle certain correlation and relevance, and humans can fit customs, habits, and even more profound causation.

In his new book, *The Book of Why*, Judea Pearl argues that machines should not only be able to connect fever to malaria but also to reason that "malaria causes fever." Once the causal framework is in place, machines can raise counter-factual questions, asking how causality will change under some intervention, allowing machines to think in probabilistic terms while humans think in probabilistic terms with purpose.

The main difference between "intelligentization" and automation is the ability to make decisions based on various information (from various areas) in highly uncertain situations, and the ability to self-learn, adaptability to "emergencies," and changing circumstances. Self-learning and self-adaptation are the abilities of a system to improve its built-in software independently (without external intervention), which is the ability to react when unexpected algorithms occur. Based on this, the following theoretical framework of the Adaptive Control of Thought-Distributed Deep Situation Awareness can be tentatively proposed:

Human-machine hybrid intelligence = biologic intelligence + nonbiologic intelligence = stimulus/choice/response + stimulus/response.

Among these, how to stimulate, choose, and respond will be the difficult and important points of construction.

The ACT-DDSA (Adaptive Control of Thought-Distributed Deep Situation Awareness) consists of database + (old and new) knowledge graph + task requirements (various environmental conditions) + situation graph + human capability (state of loading, etc.) + machine capability (machine state, etc).

The main difference between this cognitive framework and the previous ones is that it has the mechanisms of assimilation and adaptation transfer. On transfer, John Robert Anderson and others put forward the "common element theory," a new version of Edward Thorndike's theory in the psychology of information processing, in which productive rules substituted common elements. To understand how knowledge transfers across domains is critically dependent on analysis of cognitive tasks (examining the structure of knowledge acquired in one domain and assessing its applicability to another). According to the "common elements theory," the transfer between the two skills takes place under the condition that they share the same procedural knowledge, and the amount of transfer between the two skills can be evaluated by calculating the amount of procedural knowledge they share. If more procedural knowledge is shared between the two skills, there will be more significant transfer between them. If less procedural knowledge is shared between the two skills, even if they share the same declarative knowledge, there will be little or no transfer between them. This transfer may be similar to the mapping defined by randomized function. In fact, how to define it is not so critical. It is the idea that views learning just as a single-layer mapping that limits the theory. How to generate reasonable multilayer mapping (including virtual-real mapping) is the key.

The learning mechanism includes at least the following three steps: variable characterization + nonmonotonic reasoning + hybrid decision. In fact, human intelligence is not a bird, and AI is not an airplane, though many people like to make that analogy. Modern AI technology, through reasoning, knowledge, and learning, enters into a period that everyone is looking forward to, like a child full of reverie is expecting a gift box, Expectations are understandable because everyone wants to eat better after they have enough food. But this human-like intelligence lacks features of human wisdom: variable characterization + nonmonotonic reasoning + hybrid decision-making (shows common sense), and it will be difficult to satisfy the public's appetite and expectations. Human wisdom is more reasoning + more knowledge + more learning, as well as more understanding and consciousness; if understanding is to find the association, it depends on who finds it, when, where, and how to find the association. So basically, the current AI can be seen as the product under a certain situation by the programmer and staff from other fields far from the actual requirements.

A friend who recently read postmodern history for more than a year found that foreign philosophers and historians are especially good at literature, while Chinese philosophers and historians are desperately trying to go far away from literature, just like the Chinese AI scholars who are trying so hard to distance themselves from philosophy and psychology. In fact, not to mention the major of AI, even the major of computer science in established universities overseas often offer some courses in humanities, arts, history, philosophy, psychology, etc. What comes out of them are leaders surpassing the average programmers and mechanical mathematicians. Just as a famous saying goes, we lose at the starting line. Human cognitive processes show common sense, and humans are often informed and learn through stories. But AI is currently fragmented. First, some people revel in the success of machine learning, deep learning, and neural networks. They do not understand causation, and they just want to continue with the curve fitting. But when talking to people who work with AI outside statistical learning, they immediately understand. Judea Pearl makes a profound point that the inductive ideas implicit in trend extrapolation have proven unable to be a universal tool by David Hume and Karl Popper.

It's not hard to see that AI is ultimately a problem of "being" and "should." What is justice? This is also the blind point of AI. Justice means right and fair. Man himself does not know what is righteousness, so how does the machine know? As an ancient saying goes, "Jade do not become a device unless it is cut, and people do not know righteousness unless they learn." But let's think: will jade necessarily become a useful device when it is cut? If a man learns, will he know righteousness? Ren Xiangshi of the Kochi University of Technology in Japan argues that the next step in AI will not depend on traditional metrics such as whether production can be increased or certain tasks can be accomplished more quickly and efficiently, but rather on the ability to unleash human potential. It shall be designed to take into account factors such as physical and mental health, creativity, emotion, morality, and self-actualization. Human-machine hybrid intelligence can do this. Perhaps the true source of innovation does not lie in science and technology, but in the inspiration from a human's subjective world (art, literature, philosophy, religion, management). St. Jean-kévin Augustin said, "If you do not have faith, you do not need to know," which means that knowledge comes after faith. It is the same with science and religion. But for most people, they should understand first before they believe, and when it comes to the significance and value of literature, some believe that what matters most is that it opens up human possibilities. In life, we all have to weigh our options, and we are always stuck on a fixed, regular path. The American poet Robert Frost, in his poem The Road Not Taken, put it this way, "Two roads diverged in a wood, and I took the one less traveled by, and that has made all the difference." In essence, the scientific method uses homogeneous hypothesis to solve the counter-factual problem, which is often used in natural science experiments and daily life. Counter-factual relation is a virtual implication proposition, and production formula "if-then" is often used, such as love and marriage: there are no conditions when it is possible, but it is impossible when conditions permit. Such counter-facts often involve temporal stability and the transience of causality. The first scientific approach to solve counter-factual problems is to assume at the same time that time is stable, meaning the value of the output y (t) does not change with time; causality is stable, meaning the value of the output y (t) will not change whether an individual has been previously interfered with or controlled. The second scientific approach to counter-factual problems is to assume that individuals are homogeneous, that is, that two individuals are the same. In fact, however, this assumption does not exist in many cases, especially in the cognitive and social sciences.

How do relationships arise? "Being" and "should" work together. A human's "free will" can produce relationships. Both scientific skepticism and religious belief are based on the same premise: the acceptance of the invisible, and so is art. Knowledge is both universal, objective, reliable, and subjective, individualized, and fictional. We should believe as well as suspect: suspicion is scientific, and belief is religious. Knowledge is oriented toward reality, not according to the present reality. New forms of intelligence will emerge in the future when the rationality derived by machines themselves merges with human rationality (or vice versa). Cognitive scientists believe that many of the metaphors we use to understand the real world are based on how our bodies experience the physical world. The man-machine hybrid intelligence is distributed, where some are intelligent and some are nonintelligent (such as affection and meaning). People often focus on semantics and pragmatics rather than words and grammar. People's learning is not only construction, but also discovery, including not only stimulation, data, information, and knowledge, but also experience and common sense. While reading 10,000 books often involves data, information, and knowledge, traveling 10,000 miles often involves stimulating experiences and common sense. No amount of knowledge can help people swim, and no amount of theory can help people ride a bicycle. A human's situation awareness is a subjective reality, like a movie: a combination of virtual psychology and physical reality.

Adam Smith pointed out in The Wealth of Nations that "Division of labor is the starting point of civilization." The basic problem of the development strategy of human-machine hybrid intelligence under the background of rapid development of the intelligence field is to find and bring into play the comparative advantage in the cognitive division system. Comparative advantage is always a dynamic phenomenon. The disadvantage in the early stage of development will gradually become a new comparative advantage with the change of the development stage, which is a relatively common phenomenon in fast developing fields. If the relatively backward parts find the interaction pattern that suits their comparative advantage and have the ability to integrate into the architecture, these parts will start to enter the orbit of the rapid growth of intelligence. In-depth analysis will guide us to that human-machine intelligence has its own unique development pattern, which is unique in that different variables have made very different contributions to the development of intelligence, but in essence there are similarities.

The determinant growth of AI seems to be data, algorithms, and hardware advantages, but the decisive elements are breakthroughs in cognition and other disciplines. One of the characteristics of human-machine hybrid intelligence is the growing interaction between the formalization and intentionality: one pole is the many influences of complex environments, and the other is the variation of compound adaptability. "Should" is to make a kind of counter-factual inference, which is to infer counter-virtual fact from the front to the back rather than from the back to the front. It is not exactly the same as the productive if-then, more focused on subjective intention. In processing natural language, the semantics contains both the components of "being" and "should." The pragmatics also contain the component of "want" in many situations. The syntax is dominated by being. So the flexibility of grammar, semantics, and pragmatics is increasing, and objectivity is waning. Grammar, the basis of program formalization, is the basis of existing AI or automation products/systems. While there are emergencies when the system is operated through a certain grammar, there is still no subjective "should" or "want" at all, so the hybrid of human and machine intelligence becomes more and more important.

The effective criterion only exists in the individual's cognitive behavior. There is a potential correlation between double probability, the notion that subjective and objective probability events are governed by chance, and the ordered pattern that can be simulated only on the basis of coincidence. Examine the probability of these coincidences and therefore the extent to which they are allowed to be assumed to occur. Order is an accident of human rationality, and natural selection can only explain why the unfit did not survive, without explaining why the fit or the unfit appeared. An organization is a temporary collection of any forces capable of achieving a specific result, the aim of which is intentionality. The formalization is a kind of organization, and the process of knowledge visualization is the transformation between the internal psychologic knowledge and the external physical knowledge. Concept map, mind map, cognitive map, and semantic network are commonly used knowledge visualization methods. There are three key problems in knowledge visualization: What kind of knowledge should be visualized? Why do we need to visualize this knowledge? How can this knowledge be visualized?

Distributed cognition is enlightening to the formation and explicitness of an individual's mental model promoted by technology. A mental model is a kind of implicit knowledge, which transforms an individual's knowledge structure or knowledge base into external characterization, contributing to knowledge sharing and collaborative innovation. Researches in cognitive science have shown that not all cognitive tasks are suitable to and can be distributed. In specific cognitive tasks, which tasks are suitable for internal characterization and which tasks are suitable for external characterization is a practical problem. Because of individual differences, it is very difficult to describe the different characterizations of cognitive tasks. However, there is a specific logical structure under various characterizations of every cognitive task, and this abstract structure of cognitive tasks can be used as a breakthrough in the study of distributed characterization. The distributed characterization of cognitive tasks will have an important influence on the design of digital learning resources in the future.

Man's conscious action always points to a certain result, which is called "aim." Aim means "mudi" in Chinese, while "mu" refers to people's eyes, and "di" means the bull's eye. The term "mudi" is an abstract concept transferred by ancient people from the specific act of aiming at and shooting the center of the target. Jonathan emphasizes technology as a "tool in the hand of the learner" and believes that learners can acquire good thinking skills through the use of technical tools. In Pierre's words, "Intelligence is accomplished rather than possessed." The premise of distributed cognition is that individuals can also get cognition in isolated situations. But the aid of tools can improve man's cognitive efficiency (if there is no external assistance, it will be difficult for man to carry out efficient cognitive processing) and effect (if there is no external assistance, it will be difficult for man to achieve the ideal cognitive results). The external environment is mainly helpful in memorizing and computing. For example, when an individual needs to search for certain knowledge, he can make use of the carrier of external knowledge storage (such as the internet, encyclopedias, etc.) to make up for the lack of knowledge base in his mind.

3. The advantages of human wisdom: from the known to the unknown

Multiple uncertainties (in knowledge and understanding) result in uncertainty of situation awareness. Experience is reality oriented, that is, from "should" to "being" Strangeness is oriented toward reality, that is, from "being" to "should." Spontaneity is unconscious and unplanned, and consciousness is conscious and planned. Language is an important means of communication between human beings and other animals. At present, the main way of communication between human and machines is using human language; the real revolution will come when machines can produce their own language or when human machines can produce a language that communicates with each other.

The connotation of human-machine hybrid intelligence is intentionality and formalization. The minimum requirement for a mathematical proof is to understand the logical sequence of mathematical proof as a purposeful process, as Henri Poincaré put it, "something that makes the proof consistent." However, there is no such consistency in the proof of machine reasoning, which includes not only a priori, an experience, a posteriori, but also the premises and conditions of common sense. First-person has no "being" but is full of "should" as in childhood. Perhaps being a parent is the real childhood, with more "being" than "should." Maybe the real childhood of mankind starts from being a parent.

Imagination is the ability to penetrate reality, like a miner's lamp in a dark tunnel. Many definitions and concepts of mathematics in junior high school are from the beginning of the form, such as $\sqrt{a}(a \geq 0)$. Many people do not know why they should be like this when learning it, and they only know that this is the requirement, and then slowly they begin to know: the physical meaning of square root is the side length of a square; its mathematical significance is to be able to quantitatively describe the solutions of various subequations. If each equation is regarded as the relationship of numbers, then the relationship of numbers can be represented by the combination of changes in the length of the sides of squares (or polygons, or non-European surfaces), and then it shows that mathematics and graphics are fundamentally consistent. Mathematics includes both numbers and graphs. Unfortunately, AI now emphasizes numbers and ignores graphs, so it forgets the meaning of shapes. Actually, AI formally is associated not only with the number and figure, but it also should be related to intentionality of nonnumbers or nonfigures. Intelligence involves sort, characterization, acquisition, and application of relationships (including complexed or uncomplexed relationships), and mathematics is part of the relationship (or even a tiny part). AI is a small part of the intelligence, so it is ironic that AI is in full swing right now. Life is like math. What was right at the beginning may not be right now. For example, in junior high school, it must be bigger than or equal to 0. But when we have learned imaginary numbers in high school, it can be smaller than 0. There is no absolute relation in the world, only relative. Many relationships come with their own premises and conditions. The premise of "I love you" is "you are worthy of my love." The implied premise of "you are worthy of my love" is "I surrender to you," and the implied premise of "I surrender to you" is "I forget myself." Many prodigal sons often deduce nonfamily resemblance to the rationality of answers to unknown questions; many red stems often use family similarity to conclude the reliability rate of an unknown trend. Life = rationalization + rationalization + mathematical physics + nonphysicalization + rationalization + managerialization + psychologicalization + physiologicalization + physicalization + liberal arts + naturalization + … + irrationalization.

The change of machine intelligence is a kind of unchanged change, and the change of natural human intelligence is a kind of changed change. For the question of change, why is machine intelligence both variable and invariant? The answer lies in whether the object changes or the characterization of the object changes. The objects of machine intelligence are mostly characterizations of the objects, not the objects themselves. A characterization can describe how an object changes, but only a kind of unchanged change. Only when the characterization is mapped to the real object can the change of the object be the changed change.

True discovery is not a strictly logical act; it is called a daring jump. Elicitation method is to irreversibly discover, and generation method is often a form of repeatable reasoning and tracing. At present, the intelligent system is still a generative, and there is no leap to elicitation. The basic reason is that there is no nonlogical logic as a discovery relation; it is not a correlation relation. Deep situation awareness also includes the tendency to generate a second situation awareness, where the first situation awareness changes, and a random change generates a new situation awareness. It also means excellent conditioning. One problem is a rational desire (or quasi-demand). Like any other desire, it assumes that there is something to be desired: an answer. We stare at the known data, not at the data itself, but take it as clues to the unknown and as components of the unknown. As Gauss says, "I have had my answers for a long time, but I have never known how to come up with them." Gueret said, "Well … Well, we assume that everything has a soul, in personification. The beam must examine all possible paths, calculate how long each path will take, and choose the one that will take the least time." "To do so, the beam must know where it is going. If the destination is point A, the fastest route is completely different from point B." Without a clear destination, the term "fastest route" loses its meaning. Perhaps the beam had to know everything in advance, long before it set off.

4. From machine learning to learning machines

Spatial perception is the basis for understanding concepts and symbols. Only when concepts point to things or objects in space can they be meaningful and understood. The epistemology of human beings is regarded as an implicit ontology of mind. The mechanism by which machine intelligence operates can be thought of as a system of artificial knowledge without the mind. Small team deep situation awareness is organic cooperation in the form of physiologic empathy, psychological sympathy, and physical empathy. However, an isolated gorilla is not a gorilla. Basketball is a collective project involving 10 players, five for this team, five for that team, referee, audience, and TV.

Modern physics has gone through three stages, each of which has changed people's understanding of the world: from believing in a system of numbers and geometric figures, to believing in a system of masses subject to mechanical constraints, to believing in a system of mathematical constants. A fourth has yet to emerge, but the trend, such as quantum, is toward a system with subjective components. Subjective, simply speaking, is to know all the words but not being able to understand them. Just like the situation that you found a beautiful flower, but you do not know why it is beautiful. Why are flowers beautiful? First, people are used to physical symmetry, shape, color, size, etc. Of course, a few people do not like symmetry, such as some artists. Second, flowers are fragrant, arousing physical comfort. Third, flowers indicate some good characterization psychologically. In short, this is an optimization process of nonfunction. There are many levels of deep situation awareness. Huang Binghuang said that when you play golfing, you just focus on improve your own skill; when you play badminton, you have to pay attention to your partner; and when you play basketball, you have to pay attention to the cooperation between all the team players.

Learning machines from machine learning is a frog leap. Like children, they have awareness and cognitive intelligence: self-organization, self-adaptation, self-reliance, rather than it organizing, adapting, automation, reasoning amplifier, situation amplifier, situation sense amplifier, or sense amplifier. It is often exaggerated or distorted reasoning, induction, deduction that attracts people's attention, so we can try to build an awareness amplifier, a situation amplifier, or a situation awareness amplifier to solve the problem of eye-catching, which is the origin of deep situation awareness and the enhancement of attention.

The main function of formalization is to reduce the implicit factors into more restrictive and obvious informal operations, but it is absurd to completely eliminate the individual implicit participation. The most important theorem that limits the formalization of logical thinking came from Kurt Gödel. Those theorems are based on the fact that in any deductive system including arithmetic, it is possible to construct a formula, a proposition, within which no proof can be made. For example, in a film, there are often incoherent situations (including space-time, plot, emotion), discrete scenes presented by the director in front of you. The film system itself that these scenes constitute is meaningless: it becomes a meaningful story when you the audience organize these scenes together with your mind and eyes. Here again, the act of consent turns out to be logically similar to the act of discovery. Both are essentially nonformalistic, intuitive decisions of the mind.

Since the financial crisis in 2008, Mises, who once got marginal and then came to prominence after the end of the Cold War, has suddenly become more popular. And why is that? Because the Austrian economist argued that human behavior is complex, and it is with free will, and therefore difficult to calculate and plan. John Maynard Keynes understands this principle: it seems to me that scientists generally follow the basic assumptions of the properties of the laws of matter, that the system of the physical world must be composed of individuals. Individuals (situations) exert their independent and constant effects respectively. Changing in the overall state is superimposed by changes of many individuals (situations), and the individual variation is purely the result of each part of the individual's previous state. However, for a whole of different levels of complexity, there are likely to be very different laws. There will be laws of association between the complexes. However, the law of association among the complexes cannot be expressed by the law connecting individual components. The traversal process is an old routine that comes and goes, and it does not change with time or experience. In the real world, if a process is ergodic, the probability distribution of that process a thousand years later looks exactly the same as it does now. You can sample the past of the process, get a probability distribution, and predict its future. The mechanical processes that drive the physical world are ergodic, as are many biologic processes.

Not long ago, Julie Dugdale, an AI researcher at the University of Grenoble in France, began studying human behavior under stress. "In earthquakes, we find that people are more afraid of not having family or friends around than of the crisis itself," she said. The first thing people do is to look for their loved ones, and they are willing to put themselves at risk in the process. The same thing happens in fires.

5. Strong AI: a handshake between human wisdom and artificial intelligence

The term "strong AI" was originally coined by John Searle for computers and other information-processing machines. Strong AI believers think that a computer is not merely a tool for studying the human mind; on the contrary, it will own its own idea as long as the proper programs are run. But Searle himself does not believe that computers can think as people do, and in this paper, he is constantly trying to prove it. The definition he proposes here is simply what he thinks the "strong AI" researchers think, not what strong AI researchers really think. So a lot of people disagree with him.

A machine with a "strong AI" is not only a tool; it has an intelligence. A strong AI with real reasoning and problem-solving skills would be considered to have self-aware consciousness.

There are two types of strong AI.

Human-like AI means that machine thinking and reasoning is like the human mind.

Nonhuman-like AI, in which machines produce awareness and consciousness that are completely different from humans, uses reasoning in ways that are completely different from humans.

Human-machine hybrid intelligence is not a simple man + machine, but a man × machine. In short, it is to make full use of the strengths of man and machine to form a new form of intelligence. It is the result of all kinds of "bounded rationality" and "bounded sensibility" superimposing each other and surging back and forth.

Human-machine hybrid intelligence is a new type of intelligence system that is produced by the interaction of man, machine, and environment. The reason why it is different from human intelligence and AI is embodied in three aspects. First, at the intelligent input end, it combines the data objectively collected by the device sensor with the information perceived by humans subjectively, to form a new way of input. Second, it builds a unique way of understanding by integrating machine data calculation with human information cognition while processing intelligent data and information. Third, at the intelligent output end, it matches the result of machine calculation with a human's value decision, and it forms the optimized judgment from the coordination of probability and regularization. Human-machine hybrid intelligence is also a form of multiagent intelligence in a broad sense. Here, people include not only individuals but also the masses. Machines not only include equipment but also involve mechanism. In addition, it also links natural and social environments, real and virtual environments, and so on. We should focus on solving the aforementioned intelligent problems in the process of human-machine hybrid, for example, many forms of data, information characterization, a variety of logic, nonlogical reasoning, autonomous optimization decision-making, and so on.

The research of human-machine hybrid intelligence is the result of the development of intelligence technology, which includes not only the research of AI but also the exploration of the relationship between machine and man and machine and environment.

The human-machine hybrid requires defining roles and responsibilities, as well as establishing rules for human-machine hybrid collaboration. The root of this functional allocation lies in translating human needs, functions, and tactics into machine perception, capability, and execution, that is, how to integrate human perception, understanding, prediction, and feedback with machine input, processing, output, and iteration.

If man's intelligence can be divided into rational, emotional, and mental intelligence. Then the existing AI is mainly about the rational part; ethics and religion are often related to the emotional part; the mental intelligence is creative consciousness derived from humanity and art. Rational intelligence involves human experience, norms, and common knowledge; emotional intelligence includes transcendence, emotion, and belief (believing without seeing); and mental intelligence involves intuition, irrationality, and imagination. Intelligence is not a mathematical proposition of the EITHER/OR, but rather a condition and an attempt to be true or false, a flexible switching, empathy, and transference between multiple selves. A man grows from an infant to a child to a teenager to an adult, growing from being without intelligence to being with mental intelligence to being with emotional intelligence and to being with rational intelligence. It is a process from instinct to intelligence.

In short, it is safe to say that strong AI, just around the corner, is human-machine hybrid intelligence.

6. 100 questions about intelligence

(1) What is the nature of intelligence?
(2) What is the nature of stupidity?
(3) What is the nature of autonomy?
(4) What is the nature of complexity?
(5) What is the nature of wisdom?
(6) What is the nature of the characterization?
(7) What is the nature of mathematics?
(8) What is the nature of purpose?
(9) What is the nature of logic?
(10) What is the nature of motive?
(11) What is the nature of dialectics?
(12) What is the nature of analogy?
(13) What is the nature of adaptation?
(14) What is the nature of paradox?
(15) What is the nature of the human brain?
(16) What is the nature of self-ego?
(17) What is the nature of multiagent intelligence?
(18) What is the nature of feeling?
(19) What is the nature of perception?
(20) What is the nature of intuition?
(21) What is the nature of consciousness?
(22) What is the nature of cognition?
(23) What is the nature of learning?
(24) What is the nature of memory?
(25) What is the nature of interaction?
(26) What is the nature of common sense?
(27) What is the nature of understanding?
(28) What is the nature of insight?
(29) What is the nature of data?
(30) What is the nature of information?
(31) What is the nature of knowledge?
(32) What is the nature of reasoning?
(33) What is the nature of the axiom?
(34) What is the nature of the algorithm?
(35) What is the nature of decision-making?
(36) What is the nature of causation?
(37) What is the nature of explanation?

(38) What is the nature of attention?
(39) What is the nature of trust?
(40) What is the nature of emotion?
(41) What is the nature of duty?
(42) What are the limits of mathematics to the field of intelligence?
(43) What is the difference between calculation and suanji?
(44) What is the difference between mathematics and logic?
(45) What is the difference between sensibility and rationality?
(46) What is the difference between reality and virtuality?
(47) What is the difference between ethics and morality?
(48) What is the difference between human intelligence and other creatures' intelligence?
(49) What is the cognitive difference between women and men?
(50) What is the cognitive difference between children, adults, and the elderly?
(51) What is the law of transference of isomorphic and homogeneous knowledge?
(52) What is the law of transfer of isomeric and heterogeneous knowledge?
(53) What is the law of situation and nonsituation transfer?
(54) How do facts and value transform?
(55) How are intentionality and formalization transformed? How does the mechanism of "getting the meaning while discarding the form" come into being?
(56) How do state, trend, sense, and percept transform?
(57) How do axiomatic and nonaxiomatic reasoning hybridize?
(58) How do decision-making through intuition and decision-making through logic hybridize?
(59) Why is advanced intelligence a product of the interaction of human material (machine) environment systems rather than brain-like intelligence?
(60) What is the difference between multiagent intelligence and individual intelligence?
(61) What is the dynamic interface between humans, machines, and the environment?
(62) How are the different disciplines of physics, mathematics, physiology, psychology, ethics, and management transformed in the field of intelligence?

(63) Why is it that science is only part of intelligence, and intelligence is part of the field of complexity?
(64) How do computing, perception, cognition, insight, and other forms of intelligence translate into each other?
(65) How do certainty and uncertainty translate into each other?
(66) What is the difference between digitalization, automation, autonomy, "intelligentization," and intellectualization?
(67) How does one treat intelligence as an ability to deal with the contradiction between time and space effectively?
(68) In the game theory, how important is the role or function (pros and cons) played by AI at present?
(69) How does one think of the disadvantages and advantages of intelligence?
(70) How does intelligence integrate systems of different perspectives, granularity, levels, and cultures, especially between nonfamilial similarity systems?
(71) How does one deal with infinite intelligence in finite ways?
(72) How does one treat the engineering of "interconnection" and "mutuality" in intelligence?
(73) How does one treat the phenomenon of "cleverness may overreach itself" in virtual training?
(74) How does one treat the relationship between human-machine environment and grammar, semantics, and pragmatics?
(75) How do the negation, equality, and implication between humans and machines come into being?
(76) What are the differences and connections between human-machine interaction (ergonomics), human-machine mixing, human-machine hybrid, and human-machine environment system intelligence?
(77) How does empathy mechanism come into being in human communication?
(78) How does the bivergentum effect of concepts in a human's implicit and explicit knowledge come into being with the change of environment and tasks?
(79) What is the relationship between different mathematical tools in the human-machine environment system and the embedded co-location of subtasks? How does one figure out the relationship between computability and decidability?
(80) What are the mechanisms of human mistake and error formation in a given situation?

(81) What is the mechanism for counteracting intelligence?
(82) What is the system for counteracting AI?
(83) What is the relationship between natural numbers and intelligence? How does the machine define 0 and 1?
(84) What is the relationship between topology (including mathematical topology) and intelligence?
(85) What is the difference between machine learning and human learning?
(86) How does a machine produce human-like art (such as emotional intelligence like guessing riddles)?
(87) How does one realize the mechanism of machine characterization? How does one use the ambiguity of half-truth and half-falsehood, false feelings, and so on.
(88) How does one generate machine philosophy? How does a machine achieve the dialectical thinking of humanoid or superhuman beings?
(89) What is the relationship between intelligence and openness?
(90) Is there a relationship other than analogical and-or-not and their combinations?
(91) How does one exchange and match between elicitation intelligence and generation intelligence?
(92) How does one establish a collaborative framework for organizing facts and value?
(93) How does a machine achieve convergence in counter-factual reasoning or the deception of fact or value?
(94) How does one generate a human-like machine mapping, scattering, diffusion mechanism?
(95) How does a machine come up with the concept of "bravery" and so on and implement it?
(96) How does a machine establish the concept of a similar "I" and carry out "subject suspension"?
(97) How does one realize the hybrid of human "reflection" and machine "feedback"?
(98) How does one realize the parallel processing of small- and medium-sized information and big data?
(99) How does one realize the social generalization of machine intelligence?
(100) How does one achieve effective cooperation and co-evolution between man and machine?

Postscript

AI became a hot topic after AlphaGo beat human go player Lee Se-dol in 2016, especially after the launch of the National Program for the development of a new generation of AI. Books about artificial intelligence, or AI, are almost everywhere.

If human intelligence is opposed to artificial intelligence, it is easy to lead to the decline of artificial intelligence when the development of artificial intelligence is insufficient, and it is easy to lead to the fear of artificial intelligence when the development of artificial intelligence is rapid. As quoted in the book, in the 1960s, the Hungarian scholar Mikell pointed out in "*The Dimensions of Silence*" that we know far more than we can say. It's almost a common sense that from swimming, driving to face recognition, people do it intuitively, but it's hard to tell the rules or procedures behind it. No one is going to teach a child the steps to remember or recognize someone's face because this natural process of learning doesn't need to be coded. In recent years, in discussions about the future of artificial intelligence, the MIT economist David Autor has dubbed this idea "the Polanyi paradox," emphasizing the need for common sense, flexibility, and adaptability. Human intuition and judgment cannot be replaced by automated machines. Autor points out that despite the exponential growth in computing resources of all kinds and the accelerating pace of automation and intelligence since 1990, the Polanyi paradox hinders attempts by modern algorithms to replace some human skills. This argument, echoing Dreyfus's critique of symbolic AI, again leads to a debate about the limits of AI and the possibility of breaking through them.

Probably influenced by the overall thinking of relationship and integration in the East, the author in this book attempts to transcend the opposite view of intelligence, and instead and instead depicts the future of artificial intelligence as the vision of the integration of human—machine intelligence. In this cognitive framework, the author puts forward some arguments that are worthy of discussion and reflection. In my opinion, human—machine hybrid intelligence is to make full use of the advantages of human and machine to form a new form of intelligence. On the one hand, people can understand how machines see the world and make decisions effectively within the constraints of machines, which are familiar with the people, who work with them and form a tacit understanding with them.

On the other hand, effective human—machine hybrid means bringing people's thoughts and ideas to machines; the operation of the machine conforms to the personality and habits of human beings, and changes with the environment anytime and anywhere, thus forming a kind of mental body which unifies the objective data and the subjective information and realizing the effect of "human + Machine > human or machine."

Like all organic holism, these interpretations of human-machine hybrid intelligence are more or less idealistic. It is better to regard human—machine hybrid intelligence as a kind of "compound" which can better reflect the essential relationship between human and machine than human—machine hybrid intelligence, emphasizing that the science of human—machine hybrid intelligence is to study a complex system of physics and biology, or characterizing human—machine hybrid as machine—machine fusion (machine mechanism + brain mechanism), and human—human fusion (human sentiment + human reason). It is inevitable that some blueprints are larger than plans.

But from an innovation perspective, it is these seemingly unrealistic ideas that may help people explore the future of artificial intelligence in a broader space for thinking. From the reader's point of view, there are at least two points of view in the book that reveal the fundamental direction of the development of artificial intelligence. First, cognition is not calculation, and the development of artificial intelligence should shift from calculation-centered cognition to "human-centered" cognition. Human cognitive models should be introduced into artificial intelligence to enable it to achieve human like intelligence in reasoning, decision-making, memory, etc. Second, the author believes that the ethics of artificial intelligence has a long way to go; the current ethical concept of artificial intelligence is only one that humans impose on machines. The urgent task is to find out the parts that can be structurally processed in human ethics. Only in this way can it be possible to let machines learn to form their own ethical system.

No matter the future prospects are optimistic or pessimistic, artificial intelligence can be called the last invention of mankind. Faced with artificial intelligence, the holy grail of wisdom for life on Earth, even the author, who is accustomed to reflective and criticism, is hopeful that the book's blueprint will come true—the current computing-based artificial intelligence will evolve into a combination of human mind and machine intelligence.

As the American psychologist William James puts it, wisdom is the art of ignoring, making the writer stop nagging, make the reader have the leisure time to open a book, help the penetrating to think and inspiration flying, but also hope that this is a work that has a certain height, depth, and temperature at the same time.

At the same time, the book is intended to tell readers that the world is not as Pythagoras and Galileo put it, "all things are numbers" and "mathematics is the language of the universe," but that there is a wider world beyond numbers and mathematics.

Pythagoras emphasized the importance of mathematics and tried to explain everything with numbers. This school believes that numbers are the origin of all things in the universe, and the purpose of learning mathematics is to explore the mysteries of nature. A great progress in the field of philosophy and practical mathematics is to abstract the number five from the description of five apples. Although it seems normal now, this progress makes mathematical calculation possible. Philosophically, this discovery leads people to believe that numbers are the foundation of the physical world. Pythagoras transformed the myth of numbers into the mother of gods, and all things are numbers.

Realistically speaking, the root of the current field of artificial intelligence is mathematics, whether it is symbolism, connectionism, behaviorism, and mechanism. Without the mathematical system based on rules and statistics, the building of artificial intelligence will be restored to automation, even to mechanization in an instant. Why? The reason is very simple: there are machines without humans, just as there is intelligence without wisdom.

Pascal said "Man is a reed that can think"; that thought forms man's greatness. The reed that can think means that "I do not ask for space, but for the provisions of my own thought." It is of no use how much land I own. Because of space, the universe embraces me and engulfs me, like a particle; because of thought, however, I encompass the universe.

According to the current mathematical system, machines are unlikely to produce "ideas" in the foreseeable future, much less great enough to "describe the universe." Besides the reason of "the first and second theorems of incompleteness by Kurt Gödel," there is another reason that the setting of state parameters and the logical contradictions of representation, nonaxioms, and the uncertainty of intuitive reflection in a given situation make contemporary mathematics not be able to explain. And I wonder those who really and falsely thought that mathematics could do everything can find the time to stop shouting and ask themselves whether they could make warehouse full and make themselves know etiquette without the help from parents or math at home and outside, respectively.

We should appreciate the beauty of mathematics, but we should also admit that the current mathematics is still incomplete. It is hoped that AI will not decline like theology.

Hayek argues that while the facts themselves never tell us what is right, misinterpretations of the facts have the potential to change the facts and the circumstances in which we live. One man may run very fast, but he has only one arm. But if you conclude from this phenomenon that the missing arm is the reason why he's running so fast, you naturally suggest people to cut off one arm. This is what Hayek means when he says that understanding the facts changes the meaning of the facts themselves.

"The real world is like a map, a mountain map, a weather map, a building map, and so on, " Hawking writes in his book "*Grand Design*," "an individual standing at his observation point sees local reality. The higher the observation point, the more reality he can see. What we need to do is to take back the effort of trying to change someone, to respect the other person's local truthfulness, and not to ask for the approval of others, because the overlap between a map and another map is very little. Meanwhile we try to improve our observation points to see more truthfulness." Can we build a dynamic human—machine hybrid learning model that combines real-time human modeling (processing information and knowledge) with old-time machine modeling (processing data)? Then we combine the person's situation consciousness and the machine's situation awareness with the fact and the value unifies, casting a sharp sword of "feeling others and understands the difference" and building the bridge of the cohesion and gathering.

The field of human—machine hybrid intelligence is not only a scientific problem, but also a nonscientific problem. Objectively speaking, science of complexity is a false concept, for complexity is a process of multidisciplinary integration, while science is a process of "division and learning." One is the process of aggregation and convergence, while the other is the process of diffusion. One is positive and the other is negative. So its right term should be study field of complexity. Intelligence is an outstanding representative in the field of complexity research. It is not a science of "learning from different subjects" at all, but a science of multidisciplinary complexity.

The contradiction of human–machine hybrid lies in that humans diverge while machines converge, humans belong to dialectics, but machines have rules. One spreads and the other gathers, one moves and the other stay still. Moreover, we're often not dealing with one problem rather than with a whole bunch of different problems that are intertwined. Therefore, it is difficult to achieve the goal of solution by using only mathematical logic method, so it is necessary to use formal logic, dialectical logic, even nonlogic means.

The uncertainty and unexplainability of machine learning and even artificial intelligence is largely due to people finding that the induction, deduction, analogy, and other reasoning mechanisms of inventions can indeed lead to some kind of incompleteness, completeness, instability, and contradiction, and the larger the scale of the calculation is, the greater the uncertainty and unexplainability are. And the human counterfactual reasoning, antivalue reasoning, can prevent or forewarn these formal natural defects in advance from the perspective of virtual hypothesis. As a cognitive subject, the human–machine hybrid is more helpful to solve the complexity problem, but it needs to solve the problem of how to fuse under different tasks. In addition, the single integration of a person and a machine, and multipeople and multimachine group integration from the fundamental mechanism will be very different. As the old Chinese saying goes, "two heads are better than one."

One of the key issues for the rapid development of the future society is how to keep a harmonious development when the human–machine system functions at a high speed. Here, "human" refers to managers, designers, manufacturers, marketers, consumers, maintainers, and so on. "Machine" not only refers to intelligence, software, and hardware in equipment, but also refers to the mechanism of the link between the links in the industrial chain; "Environment" is involved in many areas of the "government–industry–university–research–business" collaborative environment. An ideal system of social development should achieve efficient development, precise governance, and humane care. How do you do that? It can achieve organic balance between human nature and machine nature through the interaction of situation awareness among the three systems of human–machine environment.

The epidemic is a major test of the development of countries around the world. Thankfully, China has gradually emerged from the epidemic through hard work. The "new infrastructure" currently being laid out is markedly different from the traditional infrastructure of the past, with a focus on areas of the new generation of digital technology such as 5G, artificial intelligence, and big data. Meanwhile, we hope to be able to further strengthen the coordination of human–machine environment system planning and development with these high-techs, thus making further progress and realizing the great rejuvenation of the Chinese nation.

I'd like to thank my friend Mr Han Lei. Thanks to editor Bai Lijun, Eliza Lai and linxiaofang for their efforts in editing the manuscript.

Thanks to Wang Saihan, Tao Wenxuan, He Shuhao, Han Jianyu, and Yi Tongliang for their advice and support at the beginning of the writing of this book, and thanks to Qu Xiaotong and Rohan for their great help in the writing of this book. Thanks to Bullog.cn, Pauline Wong, Ma Jiawen, Jin Xiaoyang, Wu Yu, Liu Xin, Zhuang Guangda, Hu Shaobo, He Ruiling, Guan Tianhai, Yu Qiyang, Wang Yuhu and others for their advice during the course of writing, as well as to the National Social Science Foundation for its support of the project "philosophical research on the intelligent revolution and the prospect of human deep science and technology" (project approval number: 17ZDA028), and the stimulation, awakening, and discussion of various experts and scholars. Foundation Project: Special project for Teachers of Ideological and Political Theory in Colleges and universities of the Ministry of Education Project Name: Research project on content, carrier and mechanism of Artificial intelligence ethical literacy education for college students (20JDSZK061).

The book is also a fulfillment of a promise made to a friend many years ago. And thanks to all my friends and family for their encouragement and support, especially Bai Lijun of Tsinghua University Press for his support.

Thank you for your reading and looking forward to the joint discussion. Thank you

Index

Note: 'Page numbers followed by "*f*" indicate figures and "*t*" indicate tables.'

D

E

F

G

H

I

J

K

L

M

Kolibri 360 GmbH
Ferdinand-Jühlke-Straße 7
99095 Erfurt, Deutschland
produktsicherheit@kolibri360.de